Worked Examples in METALWORKING

G. J. RICHARDSON, D. N. HAWKINS
and C. M. SELLARS

THE INSTITUTE OF METALS
London
1985

Book 369
published in May 1985 by

The Institute of Metals
1 Carlton House Terrace
London SW1Y 5DB

ISBN 0 904357 77 5

The authors

G. J. RICHARDSON, BSc, MMet, PhD, CEng, MIM

D. N. HAWKINS, BMet, PhD, CEng, MIM

C. M. SELLARS, BMet, PhD, DMet, CEng, FIM

are all in
the Department of Metallurgy,
University of Sheffield

Reproduced from paginated master typescript and illustrations prepared by the authors

Printed and made in England by The Chameleon Press Limited, London

CONTENTS

Page

SECTION 1 - ANALYSIS OF LABORATORY DATA

In the calculation of working forces for all types of metal-working processes, a knowledge of the *appropriate* flow stress for the material is essential. In general this will depend on the composition and microstructure of the material and on the strain, strain rate, temperature and deformation mode imposed by the working process. The flow stress is determined from true stress-true strain data obtained from tests, which are usually carried out in tension, compression or torsion. This first section gives examples of the derivation of these data from test results and of the methods of interpolation that may be applied.

1.1 True stress-true strain curves from tension tests

The results from a tensile test are normally presented as a force (F) - displacement ($\delta \ell$) or conventional stress (S) - conventional strain (e) curve and these must be converted to true stress (σ) - true strain (ε), where

$$S = \frac{\text{Force}}{\text{Initial Area}} = \frac{F}{A_0} \tag{1.1}$$

$$\sigma = \frac{\text{Force}}{\text{Instantaneous Area}} = \frac{F}{A} \tag{1.2}$$

$$e = \frac{\text{Change in length}}{\text{Original length}} = \frac{\delta \ell}{\ell_0} = \frac{\ell}{\ell_0} - 1 \tag{1.3}$$

$$\varepsilon = \ell n \frac{A_0}{A} \tag{1.4}$$

When deformation is *homogeneous*, additional interrelationships are obtained as a result of the constancy of volume during plastic deformation, which means that

$$A_0 \ell_0 = A \ell \tag{1.5}$$

Hence

$$\sigma = S(1 + e) \tag{1.6}$$

$$\varepsilon = \ell n \frac{\ell}{\ell_0} = \ell n(1 + e) \tag{1.7}$$

Example - Conventional stress-strain curves from room temperature tensile tests on 18/8 austenitic stainless steel (Type 304) and on a normalised low carbon steel are shown in figure 1.1. The tests were carried out at constant cross-head speed to give strain rates of $\dot{e} = 2.8 \times 10^{-3}$ and $6.7 \times 10^{-4}\ s^{-1}$ and the reductions of area at fracture were 75% and 83%, respectively.

Except for the yield, deformation is homogeneous up to the tensile strength, when necking starts. Up to the tensile strength, applying equations (1.6) and (1.7) to selected points read from the S-e curves gives:-

<u>Table 1.1</u>

	e	S N/mm^2	σ N/mm^2	ε
304	0.002	295	296	0.002
	0.10	483	531	0.0953
	0.20	561	673	0.1823
	0.30	597	776	0.2624
	0.40	608	851	0.3365
LOW C	0.0112	166	168	0.0111
	0.10	260	286	0.0953
	0.20	277	332	0.1823

These and intermediate points are plotted to give the σ-ε curves in figure 1.1.

Once necking starts, deformation is inhomogeneous and equations (1.6) and (1.7) can no longer be applied. The σ-ε curve can then only be obtained if measurements of the neck geometry are also made, which is not standard practice.

An *estimate* of the values of σ and ε at fracture may be made from the measurement of reduction of area (RA%), since the area at fracture, A_f is given by:-

$$\frac{A_f}{A_o} = \frac{100 - RA\%}{100} \qquad (1.8)$$

Hence from equation (1.4) and the fact that from equations (1.1) and (1.2) $\sigma = S\frac{A_o}{A}$, ε_f and $\sigma_{(AV)f}$ $\left[= S_f \frac{A_o}{A_f}\right]$ can be obtained. (1.9)

Table 1.2

	e_f	S_f N/mm^2	$\sigma_{(AV)f}$ N/mm^2	ε_f	Corrected σ_f N/mm^2
304	0.54	386	1544	1.386	1297 ± 23
LOW C	0.29	154	906	1.772	720 ± 18

The calculated stress $\sigma_{(AV)f}$ is influenced by the triaxial stresses in the neck and correction is required for this. For room temperature tests an estimate of this correction may be made from figure 1.2 (Tegart 1966), e.g. for the 304 stainless steel

$$\varepsilon_f - \varepsilon_n = 1.386 - 0.35 = 1.036$$

where ε_n is the true strain at the onset of necking obtained by inserting into equation (1.7) the value of conventional strain at the maximum conventional tensile stress (marked in figure 1.1). Hence $\sigma/\sigma_{(AV)} \simeq 0.84$ (± 0.015).

Corrected values of σ_f estimated in this way are given above for the two steels. Note that σ_f may also be in error because of the increase in local strain rate as necking proceeds. Even so, the values of σ_f enable low temperature flow stress curves to be extended to about 1.5 to 2 times ε_n without excessive error. At high temperatures, the high strain rate sensitivity of flow stress and the changes in microstructure as a function of strain invalidate this extrapolation procedure.

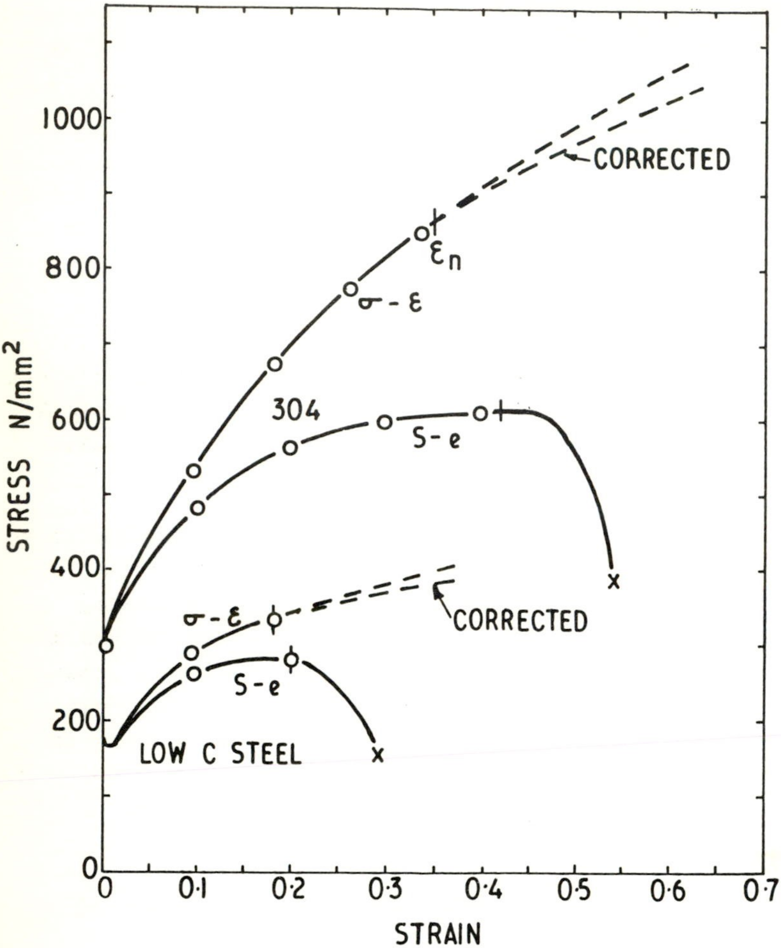
CORRECTED
ε_n
σ - ε
304
S - e
σ - ε
CORRECTED
S - e
LOW C STEEL
STRESS N/mm^2
STRAIN
0
200
400
600
800
1000
0
0·1
0·2
0·3
0·4
0·5
0·6
0·7

Fig. 1.1

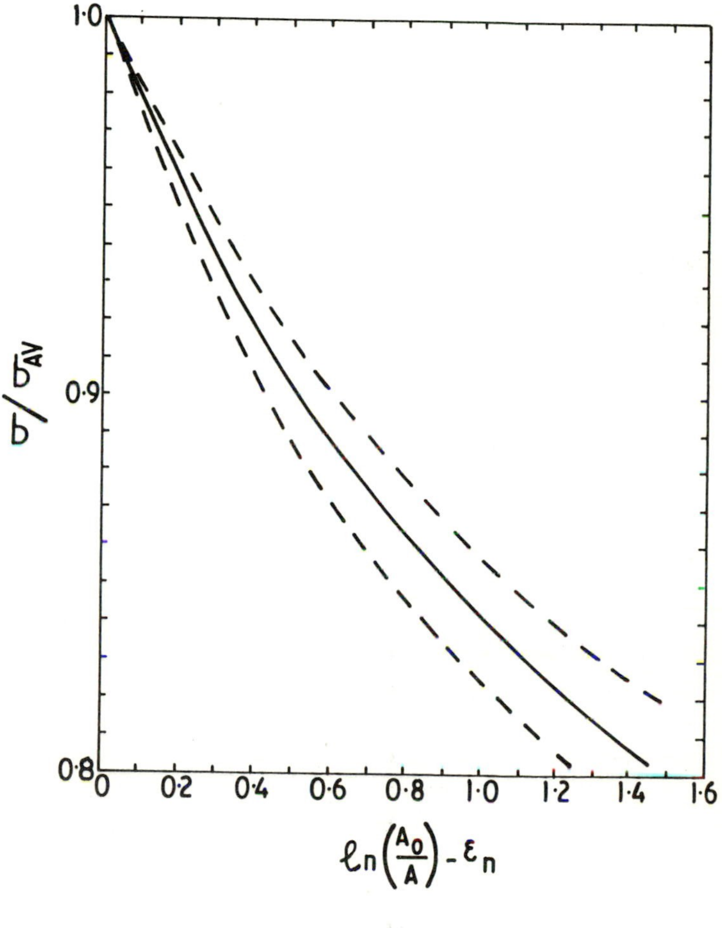

Fig. 1.2

1.2 True stress-true strain curves from axisymmetric compression tests

The problem is essentially similar to that for tension, with the additional complication created by friction between the ends of the specimen and the compression platens.

If there were no friction effects, deformation would be homogeneous and the same equations as for tension could be applied. However, *note* that when e is defined as for tension, i.e.

$$e = \frac{h}{h_o} - 1 \qquad (1.10)$$

its values in compression are *negative*. Although the negative sign is frequently omitted when strains are reported, it *must* be included in applying equations (1.6) and (1.7) to compression, e.g.

e	S N/mm^2	σ N/mm^2	ε
-0.2	37.1	29.7	-0.233
-0.5	63.2	31.6	-0.693
-0.75	117.2	29.3	-1.386

In practice the mean pressure ($\bar{p}$) on the platens is obtained from a test as

$$\bar{p} = \frac{F}{A_o}(1 + e) \qquad (1.11)$$

These values must be corrected for friction effects to obtain the compressive flow stress (σ).

If tests are carried out on a series of specimens of different initial geometries, this correction can be made graphically using the method developed by Cook and Larke (1945). When lubrication is used between the specimen ends and the platens to give a low coefficient of friction so that slipping occurs at the interfaces, consideration of the balance of forces (Rowe, 1965) leads to the algebraic relationship:

$$p = 2\sigma\left(\frac{h}{\mu d}\right)^2\left[\exp\left(\frac{\mu d}{h}\right) - 1 - \frac{\mu d}{h}\right] \qquad (1.12)$$

where μ is the effective coefficient of friction and h and d are the instantaneous values of the specimen height and diameter. This equation applies to homogeneous deformation when h and d are related to the initial values as:

$$\frac{d}{h} = \frac{d_o}{h_o}\left(\frac{h_o}{h}\right)^{3/2} = \frac{d_o}{h_o}\left(\frac{1}{1 + e}\right)^{3/2} \tag{1.13}$$

when $\frac{\mu d}{h} \lesssim 0.35$, equation (1.12) may be simplified to

$$\bar{p} = \sigma\left(1 + \frac{\mu d}{3h}\right) \tag{1.14}$$

with $\lesssim$ 1% error. If μ is known or is determined from tests on specimens of different initial geometries, these equations enable algebraic correction to be made for the effect of friction. However, they are not applicable when barrelling occurs and deformation becomes inhomogeneous or when some sticking between the specimen and platens occurs, i.e. when

$$\frac{\mu d}{h} \gtrsim \ln \frac{0.577}{\mu} \tag{1.15}$$

For these conditions only the graphical method may be used.

Example - Mean pressure-true strain curves derived from axisymmetric compression tests carried out on lead at room temperature and at constant true strain rate $\dot{\varepsilon} = 5\ s^{-1}$ are shown in figures 1.3 and 1.4. (Note that for lead room temperature ≃ 0.5 absolute melting temperature, so strain rate is critical.)

The specimens were of the same initial diameter but of different initial heights to give a series of values of $\frac{d_o}{h_o}$. One series of specimens was lubricated to give low friction conditions, $\mu = 0.04$, with the results shown in figure 1.3, and the other was unlubricated to give higher friction conditions, with the results shown in figure 1.4. Comparison of curves 1 to 4 in each of the figures clearly illustrates the importance of frictional effects. Determine the true stress-true strain curves.

Lubricated Tests - up to 50% reduction, i.e. $\varepsilon = 0.693$, no barrelling was observed. Selected values of mean pressure (p_ε) from the $\bar{p}$ - ε curves are:

Table 1.3

Curve	$\frac{d_o}{h_o}$	$\bar{p}_{0\cdot2}$	$\bar{p}_{0\cdot4}$	$\bar{p}_{0\cdot6}$	$\bar{p}_{0\cdot8}$	$\bar{p}_{1\cdot0}$	$\bar{p}_{1\cdot2}$	$\bar{p}_{1\cdot4}$
1	2.06	28.7	33.9	36.3	37.1	37.7	38.9	41.2
2	1.36	28.5	33.2	35.2	34.2	33.2	33.9	35.2
3	0.993	28.5	33.1	34.8	33.3	31.7	31.6	32.0
4	0.510	28.3	32.8	34.1	31.9	29.8	29.0	28.5

Applying equation (1.13) and then equation (1.14) to the results at strains up to 0.8 gives for $\mu = 0.04$:

Table 1.4

ε	0.2			0.4			0.6		
$\frac{d}{h}\Big/\frac{d_o}{h_o}$	1.350			1.822			2.460		
Curve	$\frac{\mu d}{h}$	$\bar{p}/\sigma$	σ	$\frac{\mu d}{h}$	$\bar{p}/\sigma$	σ	$\frac{\mu d}{h}$	$\bar{p}/\sigma$	σ
1	0.111	1.037	27.7	0.150	1.050	32.3	0.202	1.067	34.0
2	0.073	1.024	27.8	0.099	1.033	32.1	0.134	1.045	33.7
3	0.054	1.018	28.0	0.075	1.025	32.3	0.098	1.033	33.7
4	0.028	1.009	28.0	0.037	1.012	32.4	0.050	1.017	33.5

ε	0.8		
$\frac{d}{h}\Big/\frac{d_o}{h_o}$	3.320		
1	0.274	1.091	34.0
2	0.181	1.060	32.3
3	0.132	1.044	32.2
4	0.068	1.023	31.2

It can be seen that for strains of 0.2, 0.4 and 0.6 the correction is satisfactory in that a constant value of σ independent of specimen geometry is obtained within the experimental reproducibility of the curves. At a strain of 0.8, when barrelling was first apparent, the values of σ begin to show a systematic trend with geometry and this becomes worse at higher strains. The values of σ are shown as circles on the corrected curve in figure 1.3.

Lubricated and Unlubricated Tests

The graphical method can be applied to the data at all strains. It is based on the concept that the value of $\bar{p} = \sigma$ if $\frac{d_o}{h_o} = 0$, i.e. if the specimen is infinitely high so that effects arising from the ends are negligible. The method involves plotting $\bar{p}$ versus $\frac{d_o}{h_o}$ and extrapolating the curves to zero, as shown in figure 1.5 for selected values of the data given above for the lubricated tests and given below for the unlubricated tests.

Table 1.5

Curve	$\frac{d_o}{h_o}$	$\bar{p}_{0.2}$	$\bar{p}_{0.4}$	$\bar{p}_{0.6}$	$\bar{p}_{0.8}$	$\bar{p}_{1.0}$	$\bar{p}_{1.2}$	$\bar{p}_{1.4}$
1	2.05	32.6	39.7	45.0	49.0	----	----	----
2	1.37	29.9	35.5	38.2	40.4	42.9	46.0	50.0
3	0.995	29.3	34.0	36.0	36.7	37.8	39.7	42.5
4	0.499	28.7	33.2	34.3	32.1	31.3	32.0	33.7

The extrapolated values of σ obtained from the curves shown in figure 1.5 and equivalent curves for the other strains are

Table 1.6

	$\sigma_{0.2}$	$\sigma_{0.4}$	$\sigma_{0.6}$	$\sigma_{0.8}$	$\sigma_{1.0}$	$\sigma_{1.2}$	$\sigma_{1.4}$
Lubricated	28.1	32.3	33.4	30.7	28.5	27.0	26.0
Unlubricated	28.1	32.3	33.4	28.4	26.0	26.0	26.3

These values are plotted as crosses in figures 1.3 and 1.4 to obtain the corrected curves for σ.

Note that at low strains the values of σ obtained graphically for the lubricated specimens are in reasonable agreement with the values obtained algebraically. It can also be seen from figure 1.5 that the lines for strains of 0.2 and 0.4 are straight, as expected from equations 1.14 and 1.13. In fact, if μ is not known, its value may be obtained from the slopes and intercepts of these lines, or from equivalent graphs of $\bar{p}$ versus $\frac{d}{h}$.

The graphical results from the lubricated and unlubricated specimens also show excellent agreement for σ at strains $\leqslant 0.6$ but significant discrepancies arise above that. These are only partly accounted for by the increasing uncertainty in the extrapolation as the curves in figure 1.5 become steeper. This is indicated by the error bars in figures 1.3 and 1.4. The increased heterogeneity of deformation in the unlubricated specimens will also lead to differences in the distribution of temperature generated by deformation and to differences in the kinetics of the microstructural changes, which result in the reduction in flow stress at high strains. Tests on well lubricated specimens are therefore generally preferred and these are usually terminated before barrelling becomes significant, i.e. at about 50% reduction, unless the graphical method of correction is to be used.

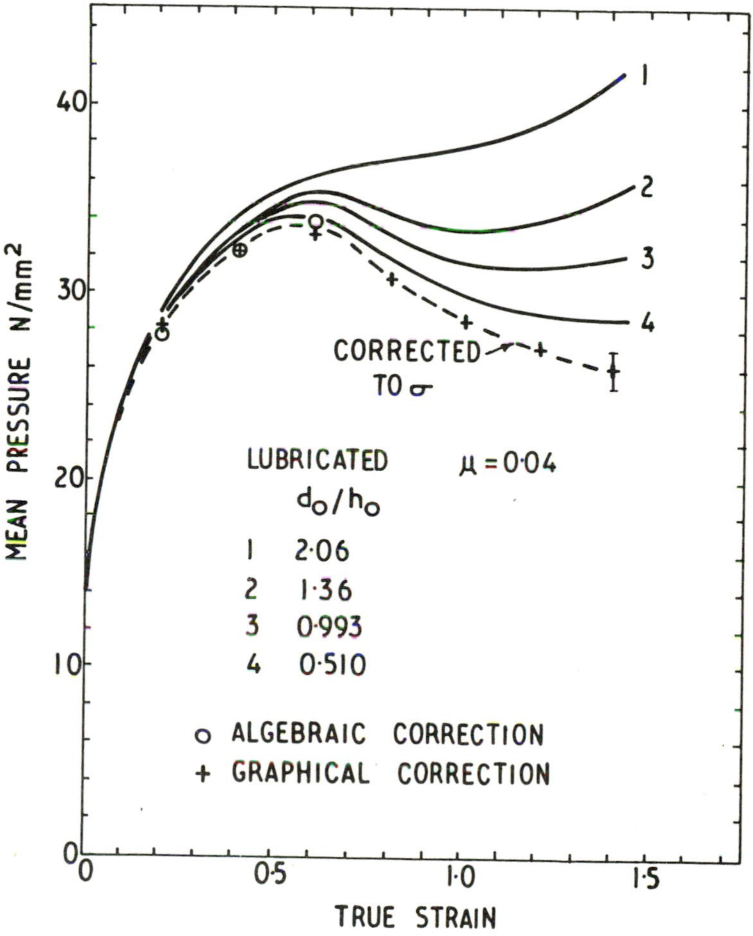
MEAN PRESSURE N/mm^2
TRUE STRAIN
0
10
20
30
40
0
0·5
1·0
1·5
1
2
3
4
CORRECTED
TO σ
LUBRICATED
μ = 0·04
d_o/h_o
1 2·06
2 1·36
3 0·993
4 0·510
o ALGEBRAIC CORRECTION
+ GRAPHICAL CORRECTION

Fig. 1.3

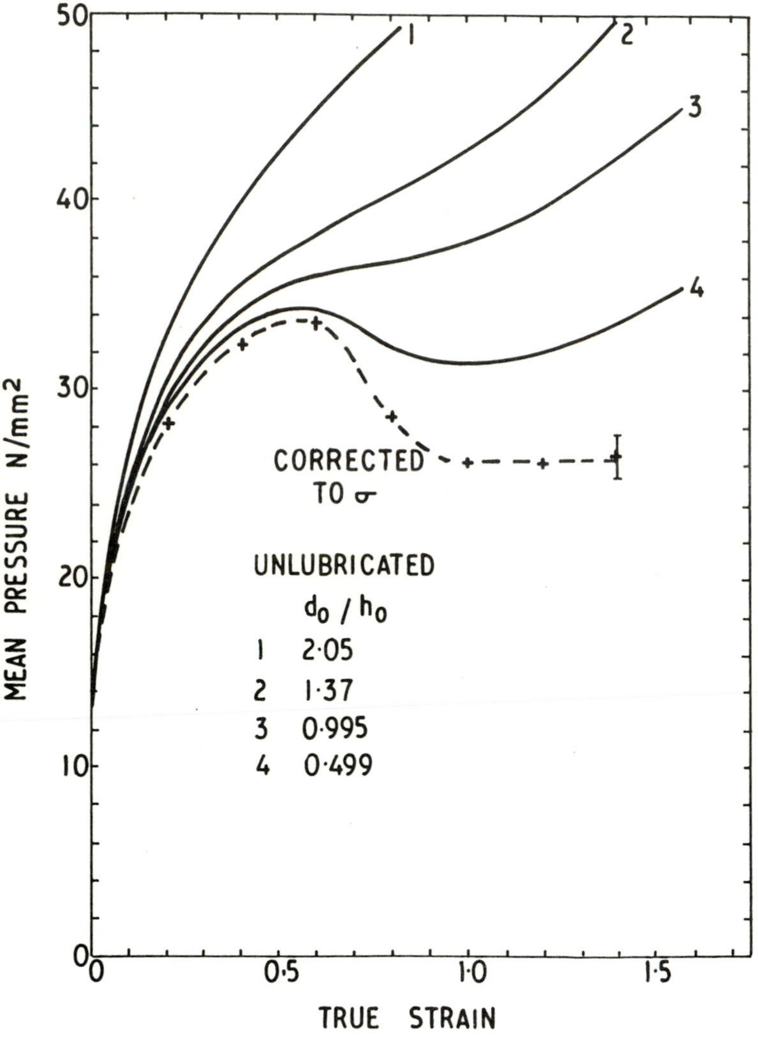

MEAN PRESSURE N/mm²
TRUE STRAIN
0
0.5
1.0
1.5
10
20
30
40
50
1
2
3
4
CORRECTED
TO σ
UNLUBRICATED
d₀ / h₀
1 2·05
2 1·37
3 0·995
4 0·499

Fig. 1.4

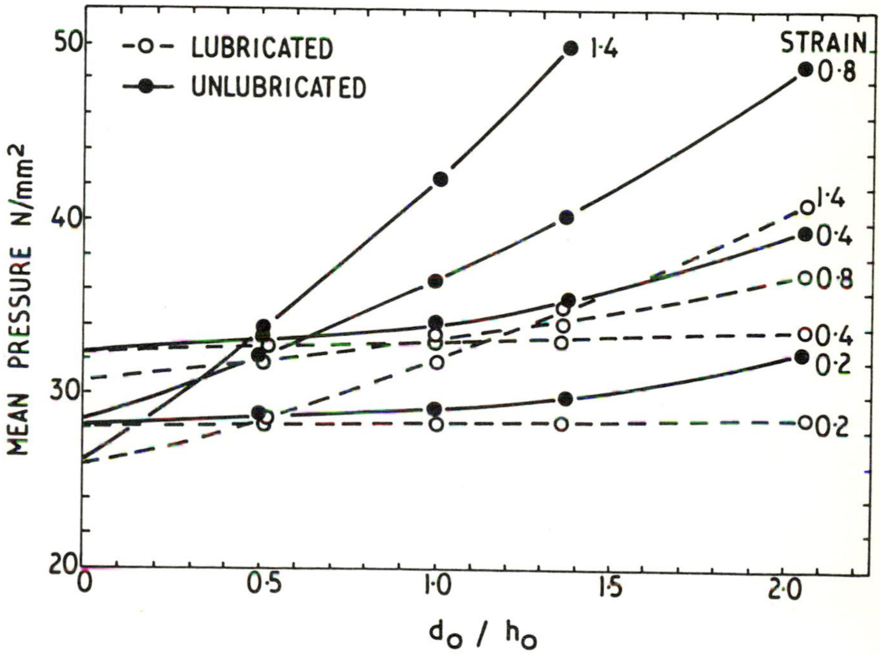

-o- LUBRICATED
—●— UNLUBRICATED
STRAIN
1·4
0·8
1·4
0·4
0·8
0·4
0·2
0·2
MEAN PRESSURE N/mm²
50
40
30
20
0
0·5
1·0
1·5
2·0
d₀ / h₀

Fig. 1.5

1.3 True stress-true strain curves from plane strain compression tests

In this test a relatively thin strip of thickness h_o is deformed between flat tools of width w_o and breadth greater than that of the strip (b_o). Normally $w_o > h_o$ and $b_o >> h_o$ so that deformation occurs in plane strain.

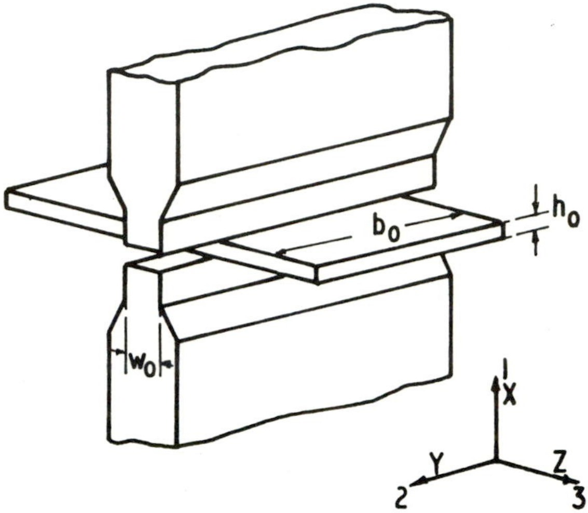

Geometry of plane strain compression testing.

In practice slight spread in breadth may take place and this must be accounted for in calculating the mean pressure from the force on the tools.

Higher strains are attainable than in axisymmetric compression tests because barrelling of the specimens does not occur. However, correction for the frictional effects between the specimen and tools must again be made even when lubrication has been used to give a low effective coefficient of friction.

In this test the reduction in height gives the *plane* strains

$$e' = \frac{h}{h_o} - 1 \tag{1.16}$$

and $$\varepsilon' = \ln \frac{h}{h_o} = \ln(1 + e') \tag{1.17}$$

where again the values of e' and ε' are *negative* although the sign is frequently omitted when the strains are reported.

From consideration of the balance of forces for sliding friction conditions (Rowe, 1965) the mean pressure on the tools ($\bar{p}$) is related to the *plane* strain flow stress (σ') as:

$$\bar{p} = \sigma' \left(\frac{h}{\mu w_o}\right) \left[\exp\left(\frac{\mu w_o}{h}\right) - 1\right] \tag{1.18}$$

whe.. h is the instantaneous thickness of the specimen and w_o is the constant tool width. Note that the plane strain flow stress is equal to twice the shear flow stress and is therefore frequently written as 2k instead of σ'.

If $\frac{\mu w_o}{h} < 0.25$, equation (1.18) may be simplified to

$$\bar{p} = \sigma' \left(1 + \frac{\mu w_o}{2h}\right) \tag{1.19}$$

with $\lesssim$ 1% error.

Equation 1.18 assumes there is no sticking between the specimens and tools and should therefore not be applied when

$$\frac{\mu w_o}{h} \gtrsim \ln \frac{1}{2\mu} \tag{1.20}$$

In fact, depending on the value of $\frac{w_o}{h}$, equation (1.18) may give deviations from the exact slip line field solutions below the critical condition given by equation (1.20) (Alexander, 1954-55).

The derived plane strain values of stress and strain may be converted to the equivalent true stress and strain by applying the von Mises criterion:

$$\sigma = \frac{\sqrt{3}}{2} \sigma' \qquad (1.21)$$

$$\varepsilon = \frac{2}{\sqrt{3}} \varepsilon' \qquad (1.22)$$

Example - A mean pressure versus plane strain curve obtained for commercial purity aluminium tested at room temperature and at a constant equivalent true strain rate of 0.6 s^{-1} is shown in figure 1.6, and selected points read from the curve are given in the table below. The specimen was initially 4.00 mm thick and was tested between tools 10.00 mm wide lubricated with a molybdenum disulphide containing graphite grease to give an effective coefficient of friction 0.02. Determine the true stress-true strain curve.

Applying equation (1.17) to the values of ε' gives the instantaneous values of h shown in the table. (Note - in practice, h would be measured in the test and ε' determined from it).

Table 1.7

ε'	$\bar{p}$ N/mm^2	h_{mm}	$\bar{p}/\sigma'$	σ'N/mm^2	ε	σ N/mm^2
0.002	59	3.99	1.025	58	0.002	50
0.05	105	3.80	1.027	102	0.058	88
0.10	122	3.62	1.028	119	0.115	103
0.20	141	3.27	1.031	137	0.231	119
0.30	152	2.96	1.035	147	0.346	127
0.50	166	2.43	1.042	159	0.577	138
0.75	184	1.89	1.055	174	0.866	151
1.00	198	1.47	1.071	185	1.155	160
1.25	213	1.15	1.092	195	1.443	169
1.50	231	0.89	1.121	206	1.732	178
1.75	249	0.70	1.157	215	2.021	186
2.00	272	0.54	1.210	225	2.310	195

The smallest value of h gives

$$\frac{\mu w_o}{h} = \frac{0.02 \times 10}{0.54} = 0.37$$

whereas $\ell n \frac{1}{2\mu} = 3.22$, so from equation (1.20)

no sticking is expected and equation (1.18) (or equation (1.19) when $h > 0.8$ mm) can be applied to obtain the values of $\bar{p}/\sigma'$ shown in the table, and hence the values of σ'. These are plotted as the corrected curve in figure 1.6.

This plane strain flow stress curve is converted to the equivalent tensile flow stress curve by applying equations (1.21) and (1.22) to obtain the values given in the table and the graph shown in figure 1.7.

Note that, although it might appear that a graphical method of solution similar to that for axisymmetric compression could be applied to plane strain results for specimens of different h_o, this would lead to large errors because of a change in mode of deformation to one of local indentation when $h_o > w_o$ (Rowe, 1965). Tests must therefore be carried out with low effective coefficients of friction if data to reasonably large strains are to be converted using the algebraic procedure given above.

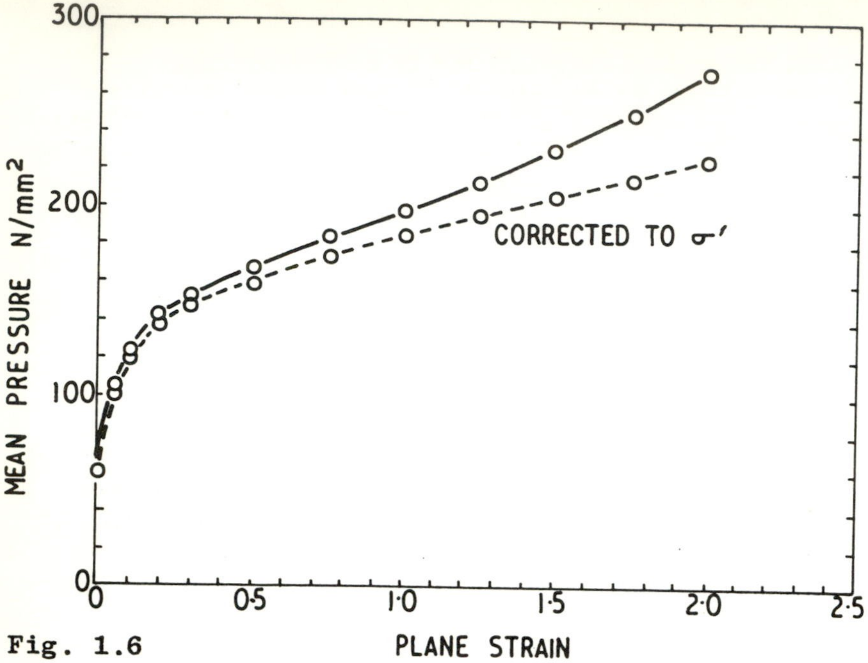
300
200
100
0
MEAN PRESSURE N/mm²
CORRECTED TO σ'
0
0·5
1·0
1·5
2·0
2·5
PLANE STRAIN

Fig. 1.6

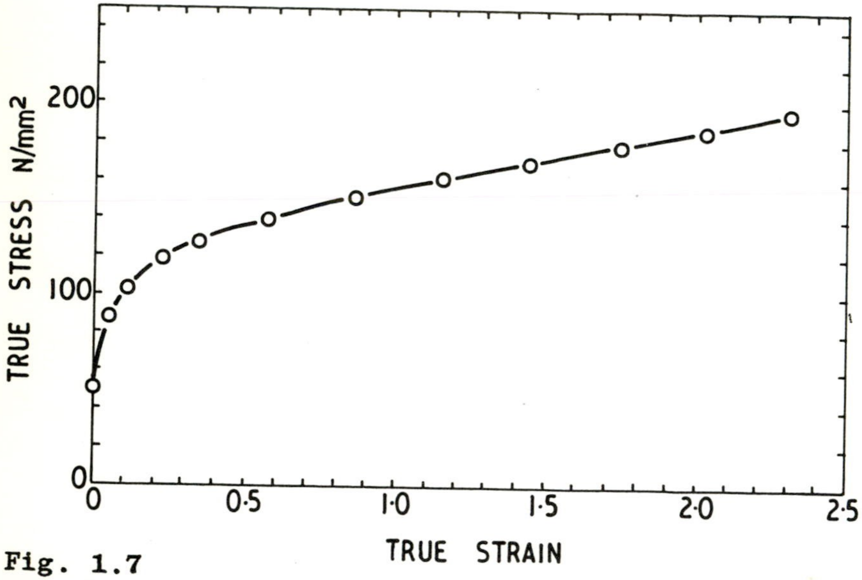
200
100
0
TRUE STRESS N/mm²
0
0·5
1·0
1·5
2·0
2·5
TRUE STRAIN

Fig. 1.7

1.4 True stress-true strain curves from torsion tests

Torsion tests for measurement of flow stress up to high strains are normally carried out on specimens with a solid cylindrical gauge section to avoid the problems of buckling of thin walled tubes. Specimens are also normally constrained longitudinally so that the gauge length (ℓ) and radius (a) remain constant during deformation. Longitudinal forces are developed because of this constraint, but these may be neglected in deriving the true stress (σ) from the torque (Γ) without significant error.

If the shear stress is a continuous function of strain and strain rate, Fields and Backofen (1957) have shown that the shear stress (τ_a) at the gauge surface is related to the torque as

$$\tau_a = \frac{1}{2\pi a^3}\left[3\Gamma + \frac{\theta d\Gamma}{d\theta} + \frac{\theta d\Gamma}{d\theta}\right] \qquad (1.23)$$

where θ is the angle of twist (radians) and $\dot{\theta}$ is the twisting speed. Alternatively, this may be expressed as

$$\tau_a = \frac{\Gamma}{2\pi a^3}\left[3 + n + m\right] \qquad (1.24)$$

where n is the slope of log Γ versus log θ at constant $\dot{\theta}$ and m is the slope of log Γ versus log $\dot{\theta}$ at constant θ. Note that the units are unimportant in plotting log-log graphs and that m is small for low temperature tests and is frequently neglected.

The surface shear strain (γ_a) and shear strain rate ($\dot{\gamma}_a$) are given by

$$\gamma_a = \frac{a\theta}{\ell} \qquad (1.25)$$

$$\dot{\gamma}_a = \frac{a\dot{\theta}}{\ell} \qquad (1.26)$$

In order to convert shear stress and shear strain to the equivalent true stress and true strain, it is usual to apply the von Mises criterion so that

$$\sigma_a = \sqrt{3}\,\tau_a \qquad (1.27)$$

$$\varepsilon_a = \gamma_a/\sqrt{3} \qquad (1.28)$$

$$\dot{\varepsilon}_a = \dot{\gamma}_a/\sqrt{3} \qquad (1.29)$$

An alternative, simpler method of converting the torque, based on the concept of an "effective radius" (a*) (Barraclough et al, 1973) is independent of m and n and gives

$$\tau_{a*} = \frac{3\Gamma}{2\pi a^3} \qquad (1.30)$$

when $$\gamma_{a*} = \frac{a*\theta}{\ell} \qquad (1.31)$$

and $$\dot{\gamma}_{a*} = \frac{a*\dot{\theta}}{\ell} \qquad (1.32)$$

Example - Torque-twist (revolutions) curves obtained from specimens of low carbon steel of gauge length 25.4 mm and gauge diameter 6.35 mm tested at 1000°C and four twisting speeds (RPM) are shown in figure 1.8 and selected values from the curves for 130 and 520 RPM are given in columns (1), (4) and (7) in the Table below. Plot true stress-true strain curves.

First Conversion Method

The equivalent true strain is simply obtained from equations (1.25) and (1.28) as

$$\varepsilon_a = \frac{3.175 \times 2\pi \times REV}{25.4 \quad \sqrt{3}} = 0.453 \times REV$$

The values are shown in column (2) below.

To convert the torque, the appropriate values of m and n must first be found. Data for the four twisting speeds are used to plot the log-log graph shown in figure 1.9. Over the relatively narrow range of twisting speeds, the slopes of the lines for a given angle of twist are approximately constant (although this may not be true over a wider range of twisting speeds). The values of m are directly obtained from the slopes of these (and other) straight lines and are shown in column (3). To obtain values of n, the torque-twist data are replotted on a log-log scale, as shown in figure 1.10 and tangents are taken at appropriate values of angle of twist. The values of n, given directly by these

Table 1.8

(1) REV	(2) ε_a	(3) m	(4) Γ Nm	(5) n	(6) σ_a N/mm^2	(7) Γ Nm	(8) n	(9) σ_a N/mm^2	(10) ε_{a*}	(11) σ_{a*} N/mm^2
0.05	0.023	0.132	2.66	0.179	71.3	3.05	0.184	87.1	0.017	78.8
0.125	0.057	0.132	3.04	"	86.7	3.48	"	99.4	0.041	89.9
0.25	0.113	0.132	3.52	"	100.4	3.98	"	113.7	0.082	102.8
0.50	0.227	0.145	4.00	"	114.5	4.52	"	129.6	0.164	116.8
0.75	0.340	0.158	4.17	0.045	115.0	4.82	"	138.7	0.246	124.5
1.00	0.453	0.173	4.17	-0.049	112.2	5.01	0.052	139.2	0.328	129.5
1.25	0.566	0.170	4.02	-0.158	104.3	5.07	0.000	138.4	0.410	131.0
1.50	0.680	0.167	3.93	-0.072	104.7	5.04	-0.082	133.9	0.492	130.2
1.75	0.793	0.165	3.91	-0.033	105.4	4.92	-0.162	127.2	0.574	127.1
2.00	0.906	0.154	3.90	"	104.9	4.75	-0.221	120.0	0.656	122.7
2.25	1.020	"		"		4.66	-0.144	120.8	0.738	120.4
2.50	1.133	"		"		4.61	-0.083	122.0	0.820	119.1
2.75	1.246	"		"		4.57	-0.058	121.9	0.903	118.1
4.00	1.812	"	3.80	"	102.1	4.51	-0.043	120.8	1.313	116.5
5.00	2.265	"	3.78	"	101.6	4.45	-0.043	119.2	1.641	115.0
			130 RPM			520 RPM			520 RPM	

tangents, are shown in columns (5) and (8) for the two twisting speeds. Note that n is approximately constant initially. This is found to be true for many materials tested both at low and high temperatures.

The equivalent true stress is now found from equations (1.24) and (1.27) as

$$\sigma_a = \frac{\sqrt{3}}{2\pi(3.175 \times 10^{-3})^3} (3 + m + n)\Gamma$$

or $$\sigma_a = 8.613 \ (3 + m + n)\Gamma$$

when σ_a is in N/mm^2 and Γ is in Nm.

The values obtained are given in columns (6) and (9) and are plotted as true stress-true strain curves in figure 1.11 for the appropriate strain rates given by equations (1.26) and

(1.29) as

$$\dot{\varepsilon}_a = \frac{3.175 \times 2\pi \times RPM}{25.4\sqrt{3}\ 60}$$

or $\dot{\varepsilon}_a(s^{-1}) = 7.56 \times 10^{-3} \times RPM$

Note that uncertainties in measuring tangents, particularly in figure 1.10, are the major sources of error, but heterogeneous deformation along the gauge length during work softening may also contribute to the minor anomalies at the beginning of steady state.

Second Conversion Method

This method does not require data for several twisting speeds and is only illustrated for the curve at 520 RPM. From equations (1.30) and (1.27)

$$\sigma_{a*} = \frac{3\sqrt{3}\ \Gamma}{2\pi(3.175 \times 10^{-3})^3}$$

or $\sigma_{a*} = 25.8\ \Gamma$ when σ is in N/mm^2 and Γ in Nm.

The values are given in column (11).

For solid specimens a* = 0.724 a (Barraclough et al, 1973) hence from equations (1.28), (1.31), and (1.32)

$$\varepsilon_{a*} = \frac{0.724 \times 3.175 \times 2\pi \times REV}{25.4\sqrt{3}} = 0.328 \times REV$$

and $$\dot{\varepsilon}_{a*} = \frac{0.724 \times 3.175 \times 2\pi \times RPM}{25.4\sqrt{3}\ 60}$$

i.e. $\dot{\varepsilon}_{a*}(s^{-1}) = 5.47 \times 10^{-3} \times RPM$

The values of ε_{a*} are shown in column (10) and are plotted against σ_{a*} in figure 1.11.

This method is simpler to apply than the first one and, as can be seen from figure 1.11, it gives a true stress-true strain curve close to that expected if the first method had been applied to a torque-twist curve for a twisting speed giving a surface strain rate of 2.85 s^{-1}. In fact numerical computation of torque-twist curves from component stress-strain curves for thin annular tubes and then conversion back by the second method indicates that this method reproduces the stress-strain curve for the annulus of radius a* with errors which are generally less than ± 1% and always less than ± 2½%. This is within the reproducibility of experimental data and is no worse than the errors introduced by the determination of m and n in the first method.

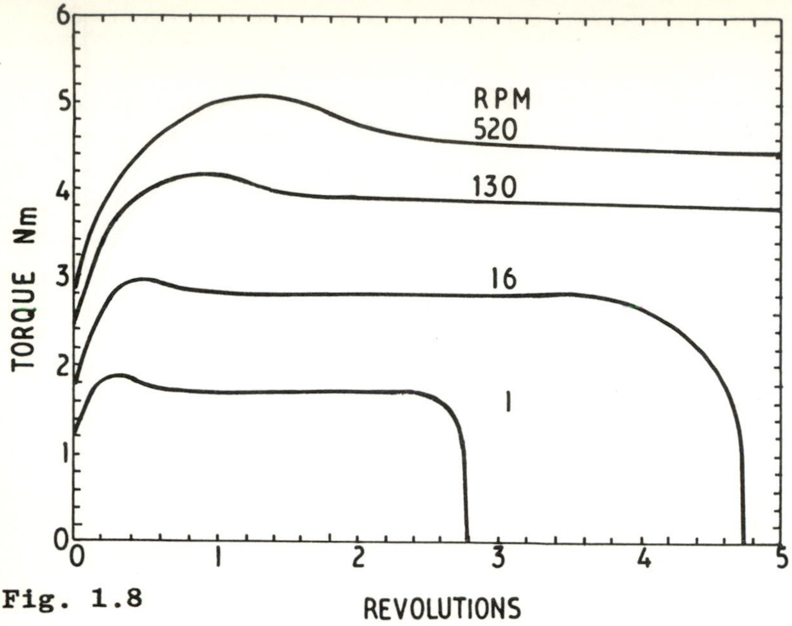
6
5
4
3
2
1
0
TORQUE Nm
RPM
520
130
16
1
0
1
2
3
4
5
REVOLUTIONS

Fig. 1.8

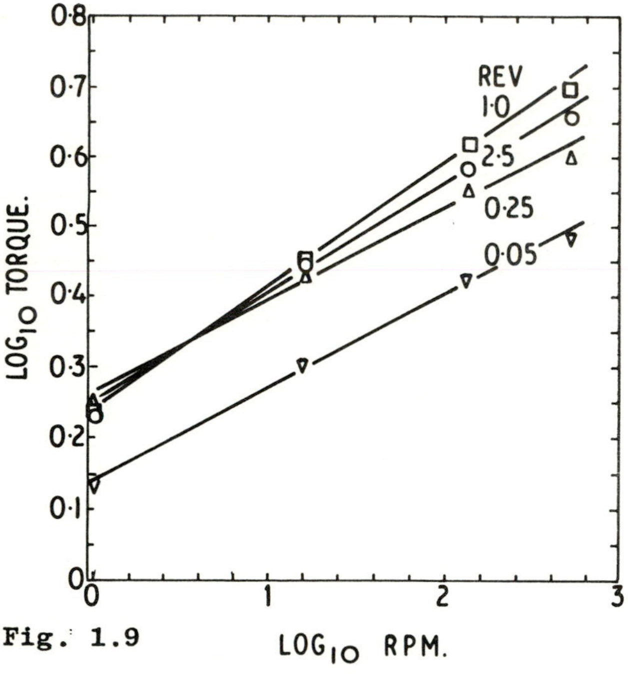
0·8
0·7
0·6
0·5
0·4
0·3
0·2
0·1
0
LOG10 TORQUE.
REV
1·0
2·5
0·25
0·05
0
1
2
3
LOG10 RPM.

Fig. 1.9

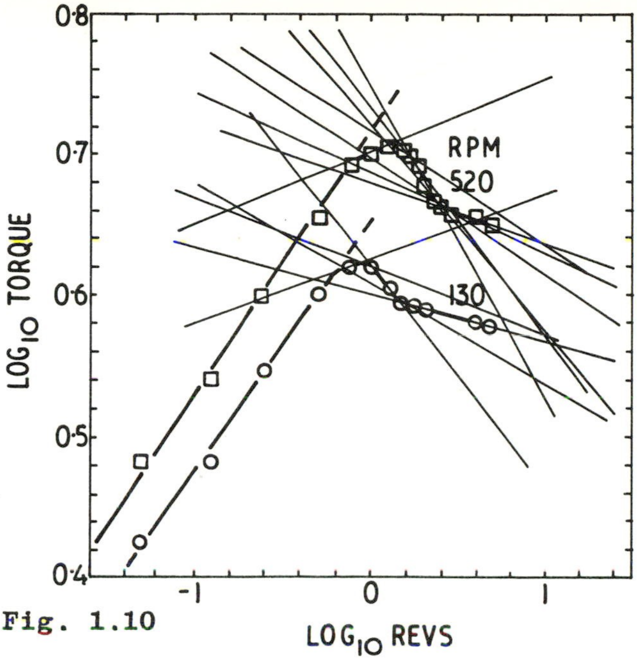

Fig. 1.10

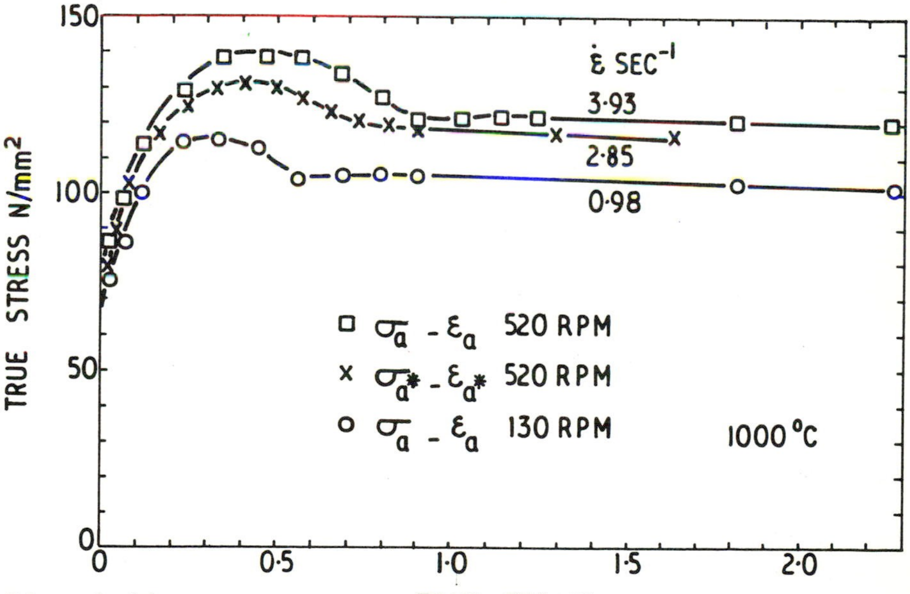

Fig. 1.11

1.5 Interpolation of High Temperature Data

The flow stress of alloys under hot working conditions is sensitive to both the temperature and strain rate of working. In a working process it is therefore necessary to consider the flow stress for the appropriate *mean* temperature and *mean* strain rate of the particular operation, e.g. a rolling pass. Frequently stress-strain data are not available for the specific combination of conditions and they must therefore be estimated from the data that are available.

For many alloys in which there is no precipitation process taking place within the temperature range of interest, the flow stress at any particular strain ($\sigma_{(\varepsilon)}$) is found to correlate with strain rate ($\dot{\varepsilon}$) and *absolute* temperature (T) as:

$$\sigma_{(\varepsilon)} = B_{(\varepsilon)}\, f(\dot{\varepsilon} \exp \frac{Q}{RT}) \quad (1.33)$$

$$= B_{(\varepsilon)}\, f(Z) \quad (1.34)$$

where $B_{(\varepsilon)}$ is a material constant, which depends on initial microstructure and strain but not on temperature and strain rate

R is the gas constant = 8.31 J/mol K

and Q is an activation energy which is often independent of temperature and in many cases is also independent of strain. Values of Q have been reported for a wide range of materials, but may be determined from data for a range of temperature *and* strain rates as shown below.

Z is the Zener-Hollomon parameter or temperature compensated strain rate.

Example - Estimate the stress-strain curve for low carbon steel deformed at a mean strain rate of 8 s^{-1} at 1067°C from the stress-strain curves given in figure 1.11 and the additional data on flow stress at the onset of steady state given in columns (1), (2) and (3) in Table 1.9.

Table 1.9

(1) Temp	(2) $\dot{\varepsilon}$ s^{-1}	(3) σ_{ss} N/mm^2	(4) $\log_{10} \dot{\varepsilon}(75)$	(5) Z s^{-1}
1100^oC	2.85	90.5	-0.09	1.16×10^{12}
(1373 K)	7.11×10^{-1}	73.0		2.89×10^{11}
	8.75×10^{-2}	48.2		3.56×10^{10}
	5.47×10^{-3}	28.6		2.23×10^{9}
1000^oC	2.85	118.9	-1.00	9.47×10^{12}
(1273 K)	7.11×10^{-1}	100.1		2.36×10^{12}
	8.75×10^{-2}	72.2		2.91×10^{11}
	5.47×10^{-3}	43.9		1.82×10^{10}
900^oC	2.85	148.6	-2.08	1.11×10^{14}
(1173 K)	7.11×10^{-1}	132.9		2.76×10^{13}
	8.75×10^{-2}	105.0		3.40×10^{12}
	5.47×10^{-3}	69.9		2.12×10^{11}

In order to use the stress-strain curves available, the strain rate and temperature dependence of flow stress must first be established. The solution therefore involves two parts:

(a) To determine the activation energy, the tabulated data are first plotted as $\log_{10} \dot{\varepsilon}$ versus flow stress as shown in figure 1.12. The strain rate required to give the *same* flow stress at each temperature can be read from this graph, e.g. for $\sigma_{ss} = 75$ N/mm^2 the values are shown in column (4) of the table.

These values and equivalent ones for $\sigma_{ss} = 50$ and 100 N/mm^2 are now plotted against $1/T$ (K^{-1}) as shown in figure 1.13. From equation (1.33), for σ = constant

$$\frac{d \log_{10}\dot{\varepsilon}}{d\ 1/T} = -\frac{Q}{\log_e 10.R} = \text{slope of graph}$$

As can be seen in figure 1.13, the slope is constant and independent of the value of stress selected. Hence, from the line for $\sigma = 75$ N/mm^2,

$$Q = -2.303 \times 8.31 \times \frac{(-2.19 + 0.12)}{(8.60 - 7.30)} \times 10^4$$

$$= 304\,735 \approx \underline{305\,000 \text{ J/mol.}}$$

Using this value of activation energy, Z may now be determined as

$$Z = \dot{\varepsilon} \exp \frac{305\ 000}{8.31\ T} \quad s^{-1}$$

to give the values shown in column (5) of the table. From these values, $\log_{10} Z$ is plotted versus σ_{ss} in figure 1.14. The data points now all fall on a single curve, which indicates that when $Z \gtrsim 3 \times 10^{11}\ s^{-1}$ (or $\sigma_{ss} \gtrsim 75\ N/mm^2$) the function in equation (1.34) is

$$Z = B'_{ss} \exp(\beta\sigma_{ss})$$

$$\text{or } \log_{10} Z = \log_{10} B'_{ss} + \frac{\beta\sigma_{ss}}{2.303}$$

with values of the constants (obtained from the slope and intercept of the graph) of $\beta = 7.67 \times 10^{-2}\ mm^2/N$ and $B'_{ss} = 1.03 \times 10^9\ s^{-1}$.

(b) This equation, or the graph, may now be used directly to interpolate values of σ_{ss} for any values of Z within the experimental range, e.g. for the conditions given in the example, $\dot{\varepsilon} = 8\ s^{-1}$, $T = 1340$ K, hence $Z = 6.29 \times 10^{12}\ s^{-1}$ and $\sigma_{ss} = 114\ N/mm^2$.

To obtain the appropriate stress-strain curve, either a series of graphs such as figure 1.14 for different $\sigma_{(\varepsilon)}$ or some complete stress-strain curves such as those in figure 1.11 are required. In figure 1.11 the three curves are appropriate to values of $Z = 1.31 \times 10^{13}$, 9.47×10^{12} and $3.26 \times 10^{12}\ s^{-1}$, i.e. the required curve for $Z = 6.29 \times 10^{12}\ s^{-1}$ lies between the lower two curves in this figure. Bearing in mind the dependence of σ on log Z in this range of Z, the best estimate of the stress at any value of strain should lie above the bottom curve by

$$\frac{\log(6.29 \times 10^{12}) - \log(3.26 \times 10^{12})}{\log(9.47 \times 10^{12}) - \log(3.26 \times 10^{12})} = 0.6$$

of the interval between these curves. The required curve can now be simply interpolated at this position.

Note that forms of temperature compensated strain rate other than the Zener Hollomon parameter are available in the literature. These are empirical but are used in essentially the same manner for interpolation purposes.

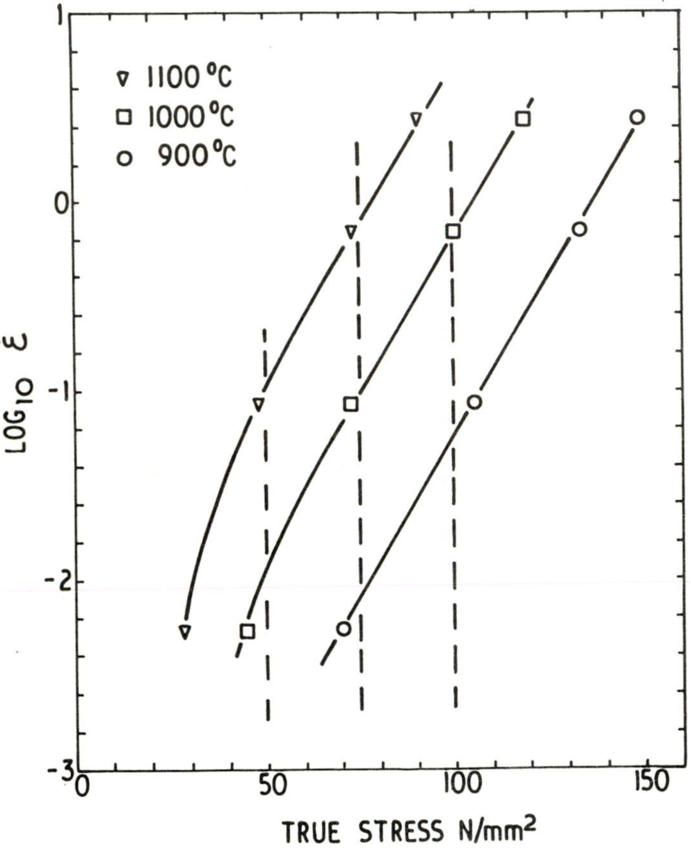

Fig. 1.12

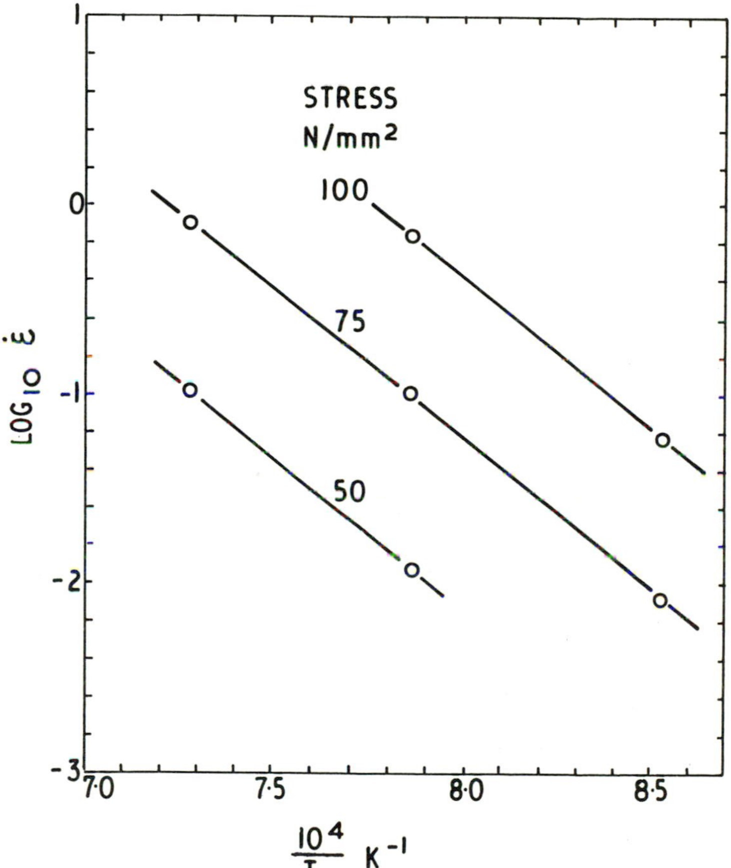

Fig. 1.13

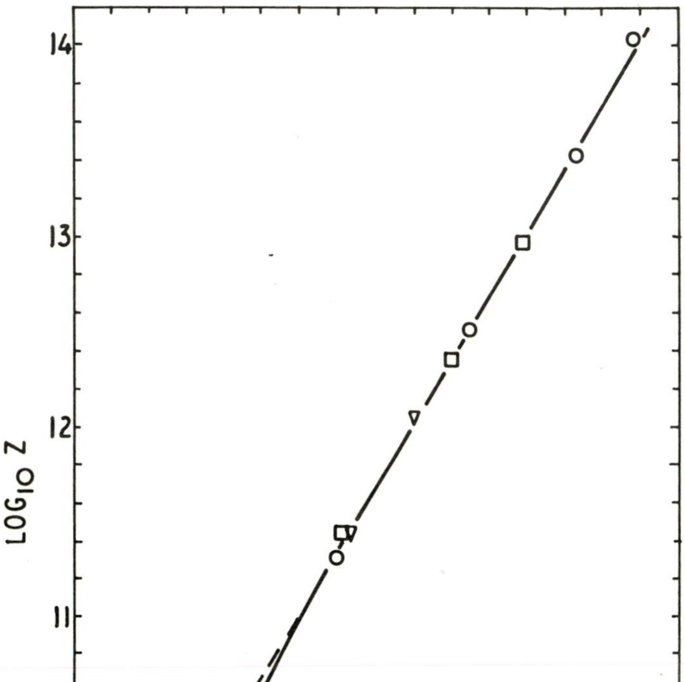

14
13
12
11
10
9
0
50
100
150
TRUE STRESS N/mm²
LOG₁₀ Z
1100 °C
1000 °C
900 °C

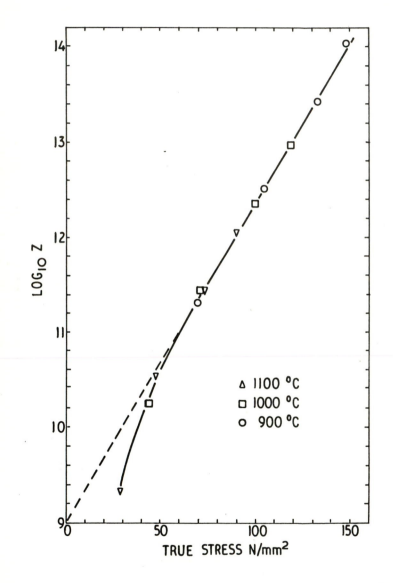

Fig. 1.14

1.6 Mean Flow Stress

The simpler theories of metal working operations assume that the metal behaves as an ideal plastic solid, i.e. that its flow stress is independent of strain. This is clearly an oversimplification, but a constant mean flow stress ($\bar{\sigma}$) can be defined from the appropriate stress-strain curve as

$$\bar{\sigma} = \frac{1}{(\varepsilon_1 - \varepsilon_0)} \int_{\varepsilon_0}^{\varepsilon_1} \sigma \, d\varepsilon \qquad (1.35)$$

Over the strain interval ($\varepsilon_1 - \varepsilon_0$) this mean flow stress gives the correct amount of work done in deformation, but clearly the value of the mean flow stress will change with the strain interval considered for a metal which is work hardening.

Example - determine the mean flow stress for wire drawing aluminium by (a) 15% reduction, (b) a second draw of 15% reduction and (c) a single draw of 30% reduction in area assuming that the appropriate stress-strain curve is given by figure 1.7 and that deformation is *homogeneous*.

From equations (1.4) and (1.8)

$$\varepsilon = \ell n \frac{100}{100 - RA\%} \qquad (1.36)$$

Thus in equation (1.35) the values of ε_0 and ε_1 are:

(a) $\varepsilon_0 = 0$ $\quad\varepsilon_1 = 0.1625$
(b) $\varepsilon_0 = 0.1625$ $\quad\varepsilon_1 = 0.3250$
(c) $\varepsilon_0 = 0$ $\quad\varepsilon_1 = 0.357$

Integration between these limits can be carried out graphically or by a numerical method, using for example Simpson's Rule. Alternatively an attempt may be made to fit the stress-strain curve by some simple relationship which can be integrated algebraically. The first and last methods are illustrated below.

Graphical Integration - this is most simply done by counting squares under the stress-strain curve in figure 1.7.

In practice the curve would be redrawn on to graph paper (from the data given in Table 1.7), say with 1 cm ≡ 20 N/mm^2 and 1 cm ≡ 0.125 (strain) so that 1 cm^2 ≡ 2.5 N/mm^2. Counting then gives:

	Area cm^2	Area N/mm^2	$\bar{\sigma}$ N/mm^2
(a)	5.92	14.80	91
(b)	7.72	19.30	119
(c)	15.26	38.15	107

Algebraic Integration - the simplest empirical equation that frequently fits stress-strain curves is

$$\sigma = K\varepsilon^n \qquad (1.37)$$

where K and n are constants. This can be checked by plotting $\log_{10} \sigma$ versus $\log_{10} \varepsilon$ as shown in figure 1.15 for the data points from figure 1.7. This gives a reasonable straight line up to $\varepsilon \simeq 1$ ($\log_{10} \varepsilon \simeq 0$). From the slope n = 0.186 and from the value of $\log_{10} \sigma = 2.185$ when $\log_{10} \varepsilon = 0$, K = 153 N/mm^2.

Over the range of strain where equation (1.37) applies, integration gives

$$\bar{\sigma} = \frac{K}{n+1} \frac{(\varepsilon_1^{n+1} - \varepsilon_0^{n+1})}{(\varepsilon_1 - \varepsilon_0)} \qquad (1.38)$$

Substitution of the appropriate values into this equation leads to

	$\bar{\sigma}$ N/mm^2
(a)	92
(b)	117
(c)	107

It can be seen that both methods give essentially the same answers. The latter method is less tedious if values of $\bar{\sigma}$ for several different strain intervals are required, but clearly it relies on the data being fitted by a simple empirical equation, whereas the former method has no such limitation. Note that whatever method is used for finding $\bar{\sigma}$, it is important, particularly for hot working conditions, that the stress-strain curve is the appropriate one for the *mean* temperature and *mean* strain rate of the working operation. As will be seen later (examples 3.2 and 3.3) somewhat different forms of mean stress may be required for specific working force calculations.

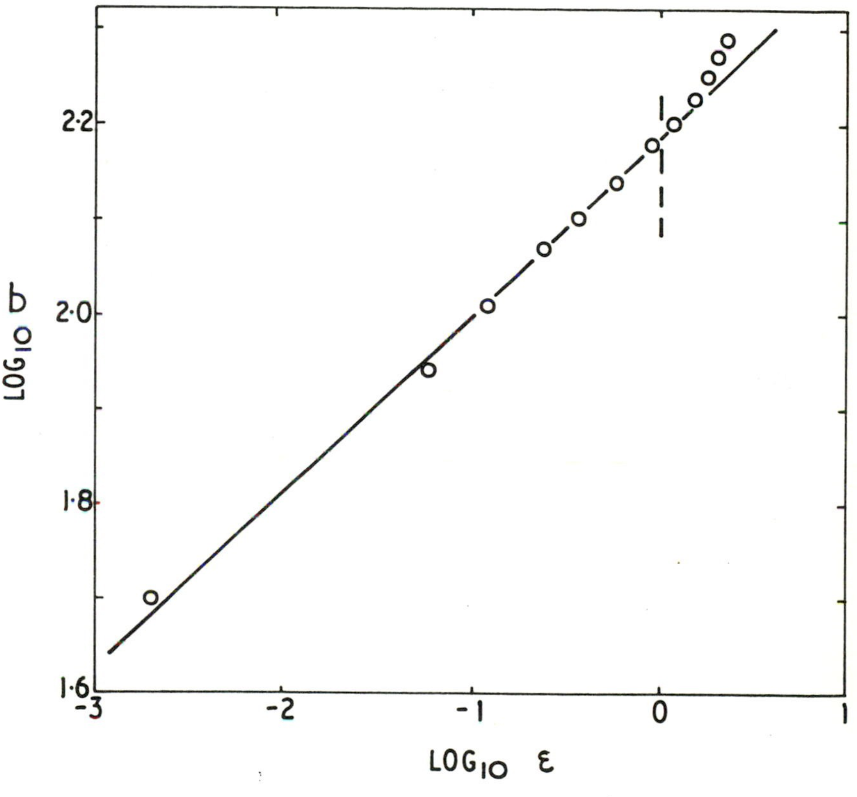

Fig. 1.15

1.7 Deformational Heating

Most of the work done during deformation is released as heat; the remainder is stored within the material as defects. The fraction (f) of the work stored depends on the strain and on the temperature of deformation; typically at temperatures low with respect to the melting temperature, it decreases with strain roughly as:

Table 1.10

ε	f
0.1	.10
0.25	.05
0.50	.025
1.00	.01

At high temperatures it approaches zero at all strains.

If the conditions of deformation were *adiabatic*, the deformational heating would result in a mean temperature rise (ΔT) given by

$$\Delta T = \frac{(1 - f)}{s\rho} \frac{W}{V} \qquad (1.39)$$

where s is specific heat per unit mass, ρ is density and W/V, the work per unit volume, is

$$\frac{W}{V} = \int_{\varepsilon_0}^{\varepsilon_1} \sigma \, d\varepsilon \qquad (1.40)$$

or, from equation (1.35)

$$\frac{W}{V} = \bar{\sigma}(\varepsilon_1 - \varepsilon_0) \qquad (1.41)$$

Example - determine the adiabatic temperature rise for the conditions given in example 1.6

$s = 900$ J/kgK, $\rho = 2700$ kg/m^3

Taking a value of f from the above table, equations (1.39) and (1.41) give for condition (a)

$$\Delta T = \frac{(1 - 0.07)}{900 \times 2700} \times 91 \times 10^6 \times 0.1625$$

$$= 5.7^{o}C$$

Similarly for conditions (b) and (c)

	f	$\Delta T^{o}C$
(b)	0.04	7.6
(c)	0.04	15.1

Note that in the calculations it is advantageous to convert all quantities to the base units of kg, N, J and m to avoid confusion.

In practice in working operations, conditions are unlikely to be adiabatic and additional inhomogenous (redundant) strains and frictional heating will contribute to the temperature rise, making accurate determination of tempera- ture a complex problem.

SECTION 2 - COLD ROLLING

The theories relating to working forces and to the interaction between the material being rolled and the equipment performing the operation are well developed and give useful and reliable results. Since the product of cold mills is required to a high degree of dimensional tolerance it is important to be able to a assess the material-equipment interaction with some confidence. The essential feature distinguishing cold rolling theories from those for hot rolling is in the assignment of the flow stress of the material. Since temperature and strain rate play only a secondary role in cold rolling, flow stress can be determined fairly accurately and the rolling theories can be expected to yield results that are close to the practical values. The main unknown parameter is the coefficient of friction, and although techniques are available for its determination, the value used under any given set of rolling conditions is usually based on empirical observation of work on other mills with similar materials. Consideration is given only to flat rolling conditions.

2.1 Cold Rolling Loads from Normal Pressure Calculations

Method (a) (Orowan 1943, Bland and Ford 1948)

A consideration of the forces exerted on a small vertical slice of strip metal as it passes through the roll gap leads to two differential equations that can each be integrated to yield the normal (vertical) pressure, p, at any point along the arc of contact (Rowe 1965).

$$p_e = \sigma' \frac{h}{h_o} \exp \mu(H_o - H) \qquad (2.1)$$

and $$p_x = \sigma' \frac{h}{h_f} \exp \mu H \qquad (2.2)$$

The two equations arise because of the change in the direction of the interfacial friction as the material accelerates through the roll gap. Equation (2.1) is applicable on the entry side of the roll gap where the rolls are travelling faster than the metal and equation (2.2) applies to the contact arc where the metal travels faster than the rolls

σ' refers to the plane strain flow stress at a particular point in the roll gap where the instantaneous thickness is h for strip that enters with thickness h_o and is reduced to h_f on exit.

μ is the effective coefficient of friction

H is given by

$$H = 2 \sqrt{\frac{R}{h_f}} \tan^{-1} \theta \sqrt{\frac{R}{h_f}} \qquad (2.3)$$

and takes the value H_o when θ, the angle subtended at the roll centre by the vertical and the point of interest, has the value θ_0 at the plane of entry.

R is the roll radius.

These equations are based on certain assumptions that relate to the flow of the metal and are well documented (Larke 1967).

(a) The arc of contact is circular.

(b) Plane strain conditions occur.

(c) Deformation is homogeneous.

(d) The coefficient of friction is constant over the arc of contact.

(e) Elastic deformation of the strip is negligible. This means that for strains in rolling of less than about 10% the equations become unreliable without correction and are not valid under temper rolling conditions (Rowe 1965).

The Rolling Load F_R.

The load is given by summing the normal roll pressure across the arc of contact and multiplying by the width of the strip.

$$\frac{F_R}{w} = \int_{\theta=0}^{\theta=\theta_0} p.R \, d\theta$$

$$F_R = R \times (\text{Area under the friction hill}) \times w \qquad (2.4)$$

This value of load will in fact require correction for the deformation of the work rolls under load before a reliable result is obtained (Section 2.3).

In addition, the derivation of equations (2.1) and (2.2) required two simplifying assumptions:

(a) the radial pressure produced by the rolls is nearly equal to the vertical pressure. For most practical strip rolling where θ and μ are relatively small the difference is only small, $\lesssim 1\%$.

(b) The change in the product $\left(1 - \frac{p}{\sigma'}\right) \frac{d}{d\theta}(h\sigma')$ is very small compared with $h\sigma' \frac{d}{d\theta}\left(1 - \frac{p}{\sigma'}\right)$.

The assumption (b) therefore precludes the equations being used for cases where the metal work hardens very rapidly, i.e. for annealed material, and where high back tensions are applied to the strip. For those conditions where the flow stress does change rapidly there are alternative treatments available that do not make these approximations (Roberts 1978).

The value of instantaneous thickness h, and the corresponding angle subtended at the roll centre θ, can be obtained from

$$h \simeq h_f + R\theta^2 \qquad (2.5(a))$$

or $$\theta \simeq \sqrt{\frac{\Delta h}{R}} \qquad (2.5(b))$$

The approximation is satisfactory since the angle of contact between entry and exit is only small and the error introduced is very small (Underwood 1950).

For the strip to be accepted by the rolls without slipping and without externally applied force the horizontal component of the frictional force at the roll surface must exceed the horizontal component of the normal roll force at the entry position. Thus for acceptance (Rowe 1965)

$$\Delta h = (h_o - h_f) \leqslant \mu^2 R \qquad (2.6)$$

Example - High purity aluminium strip 1500 mm wide is cold rolled from 4 mm to 3.3 mm thickness on a 4 high mill with 500 mm diameter work rolls. If the material has previously been reduced to a plane strain of 0.6, and the change in flow stress with strain over the range of interest is shown in figure 2.1, determine the value of the rolling load if $\mu = 0.06$.

Thus:

$w = 1500$ mm $\quad h_o = 4$ mm $\quad h_f = 3.3$ mm $\quad \Delta h = 0.7$mm

$R = 250$ mm

The calculation involves determination of the normal roll pressure at each point in the roll gap from the geometry and from the work hardening curve and summing the pressures over the arc of contact to obtain load.

The angle subtended by the arc of contact at the roll centre θ_0 is given by equation (2.5(b)).

$$\theta_0 = \sqrt{\frac{\Delta h}{R}} = \sqrt{\frac{0.7}{250}} = 0.0529 \text{ radian}$$

The constant H_o in equation (2.2) is therefore

$$H_o = 2\sqrt{\frac{R}{h_f}} \tan^{-1}\left(\theta_0\sqrt{\frac{R}{h_f}}\right) = 2\sqrt{\frac{250}{3.3}} \tan^{-1}\left(0.0529 \times \sqrt{\frac{250}{3.3}}\right)$$

$$= 7.511$$

The roll gap can be divided into a convenient number of angular intervals (Column 1 Table 2.1), the strip thickness determined at the end of each interval (Column 2), and the corresponding flow stress determined graphically from figure 2.1 for the appropriate strain at that angle (Columns 3 and 4).

The instantaneous values of θ, h, and σ' are then substituted into the pressure equations (2.1) and (2.2) to give the local value of normal pressure using each equation (Columns 8 and 11). These values are plotted against angular position in figure 2.2. Columns 5,6,7,9,10 lay out the numerical values of the intermediate terms used in the calculations.

Table 2.1

	1	2	3	4	5	6	7	8	9	10	11
	θ	h	ε'	σ'	H	exp μH	$\frac{h}{h_1}$	p_x	exp $\mu(H_0-H)$	$\frac{h}{h_0}$	p_e
EXIT	0	3.3	0.192	147.68	0	1.0	1.0	147.7	1.569	0.825	191.2
	.005	3.306	0.191	147.64	.7571	1.047	1.002	154.8	1.500	0.827	183.0
	.010	3.325	0.185	147.4	1.511	1.095	1.008	162.6	1.433	.831	175.6
	.015	3.356	0.176	147.04	2.260	1.145	1.017	171.3	1.370	.839	169.1
	.020	3.400	0.163	146.50	3.000	1.197	1.030	180.7	1.311	.850	163.2
	.025	3.456	0.146	145.84	3.730	1.251	1.047	191.0	1.255	.864	158.1
	.030	3.525	0.126	145.04	4.446	1.306	1.068	202.3	1.202	.881	153.6
	.035	3.606	0.104	144.16	5.148	1.362	1.093	216.0	1.152	.902	149.8
	.040	3.700	0.078	143.12	5.832	1.419	1.121	227.7	1.106	.925	146.4
	.045	3.806	0.050	142.00	6.499	1.477	1.153	241.9	1.063	.952	143.6
	.050	3.925	0.019	140.76	7.146	1.535	1.189	257.0	1.022	.981	141.2
ENT	.0529	4.000	0	140	7.512	1.569	1.212	266.3	1	1	140

Also shown in figure 2.2 are the curves plotted for values of μ = 0.10 and 0.14, the first of which is perhaps more typical of low (threading) speeds in the rolling of aluminium.

Equation (2.4) is now applied to obtain the total load.

Counting squares beneath the friction hill (on an enlarged diagram) gives 16 144 mm^2. But from the scale used 10000 $mm^2 \equiv 5$ N/mm^2 (x radians).

Thus the area under the curves = 8.072 N/mm^2

$$\text{Load } F_R = 250 \times 8.072 \times 1500$$

$$F_R = 3.03 \times 10^6 \text{ N} = \underline{\underline{309 \text{ tonne}}}$$

For acceptance, equation (2.6), $\Delta h < \mu^2 R$

In this case $\Delta h = 4.0 - 3.3 = 0.7$ mm

$\mu^2 R = 0.9$ mm

The metal would therefore enter the roll gap without the aid of external force.

Method (b)

A much simpler estimate of rolling load can be obtained from a modification of the equation proposed for load determination during plane strain forging between flat anvils in the presence of friction (Parkins 1968).

If the instantaneous height of the work stock, h, is replaced by the mean strip thickness $\left(\frac{h_o + h_f}{2}\right)$, and the compression length is replaced by the length of the arc of contact, $\sqrt{R\Delta h}$,

then we can write

$$F_R = w\sqrt{R\Delta h}\ \bar{\sigma}' \left(1 + \frac{\mu\sqrt{R\Delta h}}{h_o + h_f}\right) \qquad (2.7)$$

where $\Delta h = (h_o - h_f)$

w is the width

$\bar{\sigma}'$ is the mean plane strain flow stress. This may be determined graphically or algebraically from the flow stress-strain curve at the appropriate strain rate of the material.

It is defined as

$$\bar{\sigma}' = \frac{1}{\varepsilon'} \int_0^{\varepsilon'} \sigma' d\varepsilon'$$

where ε' is the plane strain at the end of the pass.

Example - Using the data in the previous section determine an approximate value of the rolling load.

Figure 2.1 shows the flow stress to be linear over the region of interest and it would thus be satisfactory to use the arithmetic mean value

$$\bar{\sigma}' = \frac{140 + 147.7}{2} = 143.85\ \text{N/mm}^2$$

However if the relationship were non-linear it would be necessary to find the area under the curve by counting methods or by algebraic integration.

From the data: $w = 1500$ mm $\Delta h = 0.7$ mm and $R = 250$ mm
and $\mu = 0.06$ $h_o = 4$ mm and $h_f = 3.3$ mm

Thus $$F_R = 1500\ \sqrt{250 \times 0.7} \times 143.85 \times \left\{1 + \frac{0.06\sqrt{250 \times 0.7}}{7.3}\right\}$$

$$= 1500 \times 13.23 \times 143.85 \times (1.109)$$

$$F_R = 3.17 \times 10^6\ \text{N} = \underline{323\ \text{tonne}}$$

Examination of the form of equation (2.7) above shows that the load is simply obtained from the product of the plane strain flow stress and the area of contact of the material with the rolls and a factor (in this case 1.109) introduced to account for the influence of friction on the rolling load. This form of equation also appears in hot rolling where the frictional factor may reach as high as four depending on the rolling geometry and on the reduction (see Section 3).

Method (c)

A third method that has gained wide acceptance, because it gives results that are useful over a wide range of rolling conditions (Rowe 1965), was proposed by Ekelund (1933).

$$\frac{F_R}{w} = \bar{\sigma}'\sqrt{R\Delta h}\left\{1 + \frac{1.6\,\mu\sqrt{R\Delta h} - 1.2\,\Delta h}{h_o + h_f}\right\} \qquad (2.8)$$

Strictly speaking the radius R in the formula should be the true (deformed) radius at the rolling interface (see Section 2.3)

Example - Using the previous data find the rolling load.

$$F_R = 143.85 \times 1500\ \sqrt{250 \times 0.7}\ \left\{1 + \frac{1.6 \times 0.06\sqrt{250 \times 0.7} - 1.2 \times 0.7}{4 + 3.3}\right\}$$

$$= 3.02 \times 10^6\ \text{N}$$
$$= \underline{308\ \text{tonne}}$$

Although the agreement between the Ekelund, the Orowan and Bland and Ford approach is perhaps fortuitous, it does illustrate how convenient Method (c) is compared to Method (a). The much less rigorous approach in Method (b) also gives a rolling load that only differs from the others by less than 5%.

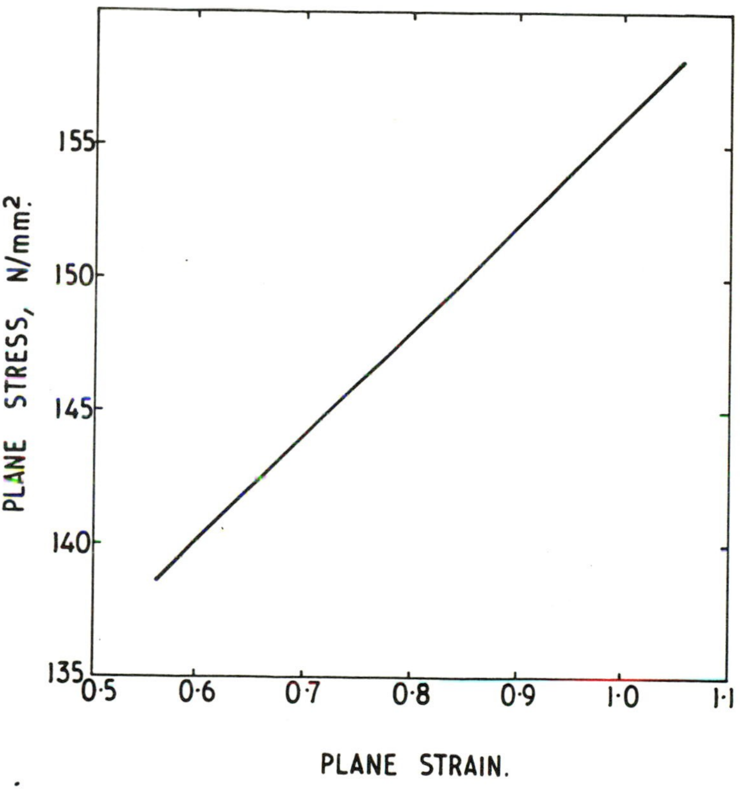

Fig. 2.1

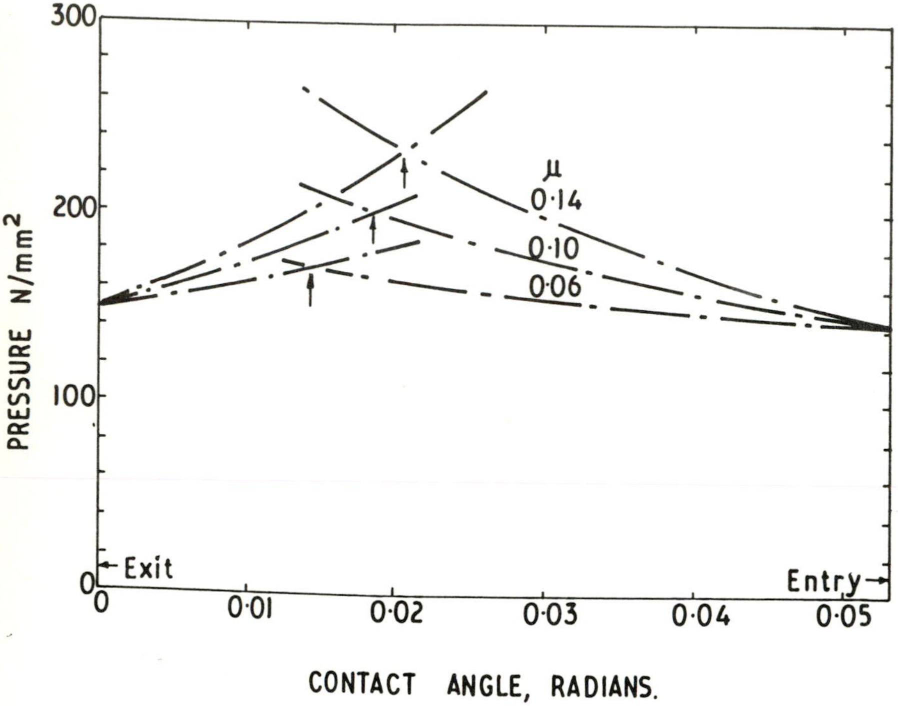

300
200
100
0
PRESSURE N/mm^2
μ
0·14
0·10
0·06
Exit
Entry
0
0·01
0·02
0·03
0·04
0·05
CONTACT ANGLE, RADIANS.

Fig. 2.2

2.2 Position of the Neutral Plane

As strip metal passes through the roll gap its thickness decreases and therefore the length of stock increases. Thus on the entry side of the roll gap the rolls travel faster than the stock and on the exit side the stock moves faster than the rolls. The position at which the rolls and the stock are travelling at the same speed defines the position of the neutral plane. This is also the position at which the direction of the frictional forces changes and thus friction is instantaneously zero.

(a) Since equations (2.1) and (2.2) describe the stresses on either side of the neutral plane, the position where the curves intersect each other defines the angle at which the neutral plane occurs. This plane can thus be found graphically, or analytically by solving these equations simultaneously.

At the neutral plane

$$\frac{\exp \mu H_n}{h_f} = \frac{\exp \mu(H_o - H_n)}{h_o}$$

and rearranging:

$$H_n = \frac{H_o}{2} - \frac{1}{2\mu} \ln \frac{h_o}{h_f} \qquad (2.9)$$

where H_n is the value of H at the neutral plane which lies at angle θ_n to the vertical line between roll centres. Thus:

$$H_n = 2\sqrt{\frac{R}{h_f}} \tan^{-1}\left(\theta_n\sqrt{\frac{R}{h_f}}\right) \qquad (2.10)$$

(b) An alternative to these rather cumbersome equations can be used to give the angle θ_n in terms of the angle of bite θ_o and the frictional coefficient.

$$\sin \theta_n = \frac{\sin \theta_o}{2} - \frac{\sin^2\left(\frac{\theta_o}{2}\right)}{\mu} \qquad (2.11a)$$

This equation is derived on the assumption that the normal roll pressure is constant across the arc of contact and that the sum of the components of horizontal force balance at the neutral plane.

It may be further simplified without serious error (Underwood 1950) to give

$$\theta_n \approx \sqrt{\frac{\Delta h}{4R}} - \frac{1}{\mu}\frac{\Delta h}{4R} \qquad (2.11b)$$

assuming that $R\sin\theta_o \approx \sqrt{R\Delta h}$.

It should be noted from equations (2.9 and 2.11) that the neutral plane is dependent solely on the rolling geometry and the friction coefficient, totally independent of the flow stress of the material.

<u>Example</u> - Using the data in Section (2.1a) determine the position of the neutral plane by each method.

(a) From figure 2.2

Intersection of the two lines occurs at $\theta_n = 0.0144$ rad

(b) From equation (2.9), using the already calculated value of $H_o = 7.511$,

$$H_n = \frac{7.511}{2} - \frac{1}{2 \times 0.06} \ln \frac{4}{3.3}$$

$$\therefore \quad H_n = 2.1524$$

Rearranging equation (2.10) gives

$$\theta_n = \frac{1}{\sqrt{\frac{R}{h_f}}} \tan \frac{H_n}{2\sqrt{\frac{R}{h_f}}}$$

$$= \sqrt{\frac{3.3}{250}} \tan \frac{2.1524}{2} \sqrt{\frac{3.3}{250}}$$

$$\theta_n = 0.0141 \text{ rad}$$

(c) From equation (2.11b)

$$\theta_n = \sqrt{\frac{0.7}{2 \times 500}} - \frac{0.7}{0.06 \times 2 \times 500}$$

$$= 0.02646 - 0.01167$$

$$\theta_n = 0.0148 \text{ rad}$$

Comparison of the three results shows that there is little error introduced by the less rigorous method used in (c) and it is much more convenient to use.

2.3 Rolling Loads Corrected for Roll Flattening

When the strip passes through the roll gap the roll separating force acts on the rolls, deforming them elastically. Although there is some evidence that the flattened rolls may no longer be circular along the contact area, it is convenient to make the assumption that they are since this simplifies the analysis. It is assumed therefore that the rolls are deformed to a new but larger radius R′ along the contact arc. Elasticity theory then allows this radius to be computed for a particular value of rolling load (Hitchcock 1935)

$$R' = R\left\{1 + \frac{CF_{R'}}{w\Delta h}\right\} \qquad (2.12)$$

where $F_{R'}$ is the true rolling load that is generated i.e. is the load based on radius R′, since this then refers to the true arc of contact.

C is a constant for the particular roll materials involved and depends *only* on the elastic constants

$$C = \frac{16(1 - \nu^2)}{\pi E} \qquad (2.13)$$

where ν is Poisson's ratio and E is Young's modulus.

(a) It is clear therefore that if the rolling load is being calculated, then the method outlined in Section 2.1 may be significantly in error since the calculations were based on radius R. The way to proceed in this case is to put the best value of F that can be calculated into equation (2.12) and determine the improved value of radius i.e. R′. With this value a more reliable value of the rolling load $F_{R'}$ can be obtained which can again be used to determine another value of radius R″ . By this iterative process a gradual approach to the true rolling load can be made. The point at which the iteration is terminated depends on the degree of accuracy required. However, too long an iteration is not justified because of the uncertainties in the other assumptions, such as the circular nature of the contact arc of contact.

Example - Based on the previous determination of rolling load Section (2.1) determine an improved value and decide when the iteration is worth terminating. Assume that the rolls are forged steel.

From the previous calculation:

$w = 1500$ mm $\quad h_o = 4$ mm $\quad h_1 = 3.3$ mm $\quad R = 250$ mm

$\mu = 0.06$ and the value of rolling load from the friction hill is:

3.03×10^6 N.

From equation (2.13):

the value of C can be determined using realistic values of the elastic constant for cold rolling rolls. Assuming that Young's modulus is 316×10^9 N/mm^2 and Poisson's ratio is 0.31 then:-

$$C = \frac{16(1 - 0.31^2)}{\pi\ 216 \times 10^9}\ m^2/N$$

$$C = 2.13 \times 10^{-11}\ m^2/N$$

Substituting the data into equation (2.12) gives

$$R' = 0.250\left\{1 + \frac{2.13 \times 10^{-11} \times 3.03 \times 10^6}{1.5 \times 0.7 \times 10^{-3}}\right\}$$

$$\underline{R' = 0.2654\ m} \quad (= 265.4\ mm)$$

This now allows us to redetermine the rolling load from the more accurate value of the arc of contact.

The new angle of contact θ_o required in the equation (2.3) is

$$\theta_o = \sqrt{\frac{\Delta h}{R'}} = \frac{0.7}{265.4} = 0.0514\ rad.$$

Similarly the new value of H_o for use in the pressure equation (2.1) is

$$H_o = 2\sqrt{\frac{R'}{h_f}}\ \tan^{-1}\ \theta_o\sqrt{\frac{R'}{h_f}}$$

$$= 2\sqrt{\frac{265.4}{3.3}}\ \tan^{-1}\ 0.0514\sqrt{\frac{265.4}{3.3}}$$

$$= 17.936\ \tan^{-1}\ (0.4610)$$

$$= 7.748$$

Finally the instantaneous thickness at any angle θ in the roll gap is now, from equation (2.5)

$$h = h_f + R'\theta^2$$

It should also be remembered that even though the arc of contact is again divided into the same increments as previously, there will be a different value of flow stress at each value of angle. In this case it is a negligible difference but it could be significant where the metal work hardens more rapidly.

Proceeding as previously, Section (2.1), gives the values in Table 2.2 and figure 2.3. The area under the new friction hill is found to be 7.976 N/mm², figure (2.3), and therefore the improved load

$$F_{R'} = 265.4 \times 7.976 \times 1500$$
$$= 3.18 \times 10^6 \text{ N}$$

or $F_{R'}$ = 324 tonne

Table 2.2

	1	2	3	4	5	6	7	8	9	10	11
	θ	h	ε	σ'	H	$\exp \mu H$	$\frac{h}{h_f}$	p_x	$\exp \mu(H_0 - H)$	$\frac{h}{h_0}$	p_e
EXIT	0	3.3	0.192	147.68	0	1	1	147.68	1.592	0.825	193.93
	0.005	3.307	0.190	147.60	.804	1.049	1.002	155.22	1.517	0.827	185.09
	.01	3.327	0.184	147.36	1.604	1.101	1.008	163.58	1.446	0.832	177.19
	.015	3.360	0.174	146.96	2.398	1.155	1.018	172.79	1.378	0.840	170.16
	.020	3.406	0.161	146.44	3.183	1.210	1.032	182.95	1.315	0.852	163.97
	.025	3.466	0.143	145.72	3.956	1.268	1.050	194.05	1.255	0.867	158.52
	.030	3.539	0.123	144.92	4.714	1.327	1.072	206.22	1.200	0.885	153.81
	.035	3.625	0.098	143.92	5.455	1.387	1.099	219.31	1.147	0.906	149.66
	.040	3.725	0.071	142.84	6.178	1.449	1.129	233.58	1.099	0.931	146.15
	.045	3.837	0.042	141.68	6.880	1.511	1.163	248.92	1.053	0.959	143.17
	.050	3.964	0.009	140.36	7.560	1.574	1.201	265.38	1.011	0.991	140.66
ENT	.0514	4.0	0	140	7.747	1.592	1.212	270.11	1	1	140

There is a slight anomaly arising in this technique, in that it is assumed that the exit plane of the stock lies along the line of roll centres. However, the classical view of roll flattening is that the exit plane lies even further out on the exit side of the rolls (Underwood 1950). This difference should lead to a slight error in calculation of load but it should be small since the arc of contact is the correct length. It is likely that this can lead to a larger error when it comes to determination of roll torque.

This improved load value can now be inserted for $F_{R'}$ into the roll flattening equation and an even better value of radius determined

$$R'' = 0.250 \left\{1 + \frac{2.13 \times 10^{-11} \times 3.18 \times 10^{6}}{1.50 \times 0.7 \times 10^{-3}}\right\}$$

$$\therefore \quad \underline{R'' = 266.1 \text{ mm}}$$

(Note that in the calculation R = 0.25 m is used once again in the equation - not R').

To determine whether it is necessary to recalculate the load the two values of R' and R'' may be compared.

$$\frac{R''}{R'} = \frac{266.1}{265.4} = 1.003$$

Thus the radius has changed by only 3 parts in 1000 after recalculation so that a re-evaluation of the load would not improve the result significantly.

Reference to figure (2.3) shows that the correction for flattening has virtually no effect on the friction hill and most of the load increase is due to the increased length of the arc of contact. Although the contact angle is smaller, the arc length is greater due to the increased radius R'.

Equation (2.4) indicates that if the areas under the friction hill are assumed to be the same then the ratio of the corrected load to the uncorrected load is in the ratio of the corrected radius to the uncorrected radius. Thus $F_{R'}$

in the example becomes 328 tonne. This strongly suggests that the recalculation of the friction hill is an unnecessary refinement; the ratio of the two radii giving adequate correction for many calculations.

The load determined from the Ekelund equation (2.8) yields a value that differs from 324 tonnes by less than two percent.

Thus:

$$F_{R'} = 143.85 \text{x} 1500 \text{x} \sqrt{265.4 \text{x} 0.7} \left\{ 1 + \frac{1.6 \text{x} 0.06 \text{x} \sqrt{265.4 \text{x} 0.7} - 1.2 \text{x} 0.7}{4 + 3.3} \right\}$$

$$F_{R'} = 3.13 \times 10^6 \text{ N} = 319 \text{ tonne}$$

This magnitude of the difference between the friction hill and the Ekelund method approaches the error that arises from uncertainties in the assumptions made and in the value assigned to the frictional coefficient μ. For example, if μ is taken as 0.066 then $F_{R'}$ is 324.4 tonne; giving a larger change in load than the change due to the method of calculation. If μ is taken as 0.12 the load becomes 372.8 tonne.

In all rolling processes there is significant doubt about the value of average frictional coefficient and it can vary by a factor of two depending on the rolling conditions of speed, type of feedstock, surface quality of the rolls and type of lubricant used. In addition it varies across the arc of contact (Underwood 1950). The value used in calculation, therefore, is the best one based on experiment and experience, but it must still be open to question. In view of these uncertainties, unless very accurate work is required and precautions taken to ensure minimum error in the parameters used, it is doubtful whether the use of the more rigorous analysis is necessary.

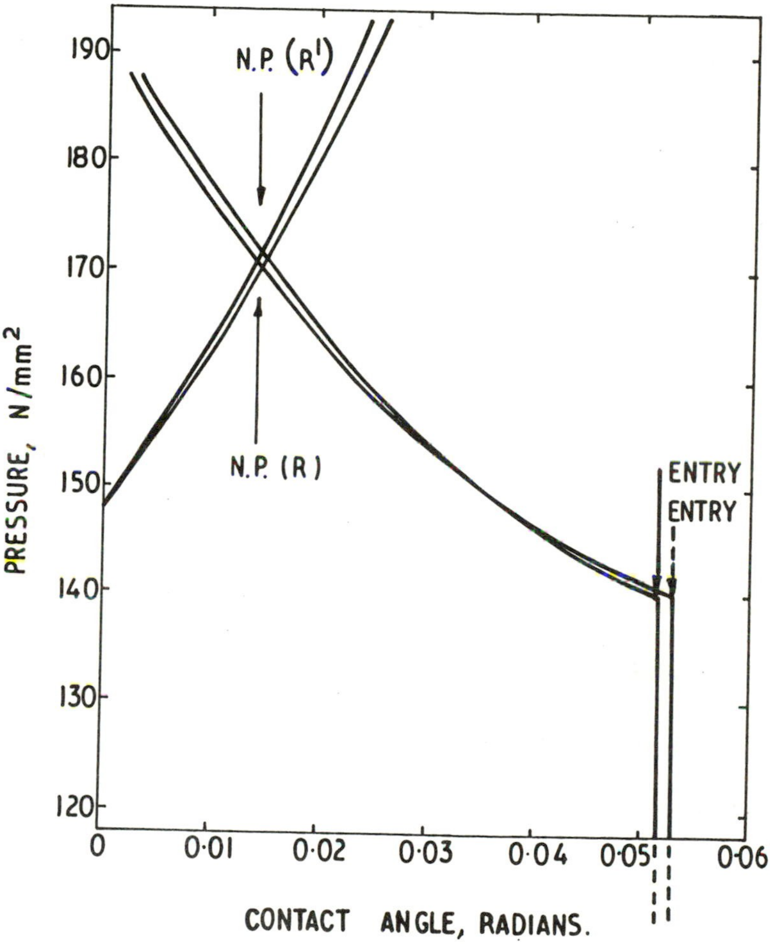
190
180
170
160
150
140
130
120
N.P. (R')
N.P. (R)
ENTRY
ENTRY
PRESSURE, N/mm2
0
0·01
0·02
0·03
0·04
0·05
0·06
CONTACT ANGLE, RADIANS.

Fig. 2.3

2.4 Coiler and Decoiler Tension

Application of front and back tension to cold strip enables the production of good gauge. However it has the secondary effect of reducing the rolling load and this can be beneficial to the rolling of very thin gauge material through the reduction of roll flattening (Rowe 1965).

The effect of tensions on normal roll pressure within the roll gap is very simply calculated once the normal pressure is known in the absence of tension (Sections 2.1 and 2.3). The pressure equations (2.1) and (2.2) are, respectively, multiplied by the factor

$$\left(1 - \frac{t_o}{\sigma_o'}\right) \quad \text{for back tension } t_o\text{, with the equation for strip } \textit{entry} \tag{2.14}$$

and by

$$\left(1 - \frac{t_F}{\sigma_F'}\right) \quad \text{for front tension } t_F\text{, with the equation for strip } \textit{exit} \tag{2.15}$$

These factors arise from the insertion of horizontal stresses as boundary conditions in the derivation of the rolling equations. The values of σ' to be used in the above equations should therefore be the *ingoing* flow stress at entry, σ_o', and the *final* flow stress at the exit, σ_F'.

The equations (2.1) and (2.2) thus become

$$p_x = \left(1 - \frac{t_F}{\sigma_F'}\right) \sigma' \frac{h}{h_F} \exp \mu H \tag{2.16}$$

and

$$p_e = \left(1 - \frac{t_o}{\sigma_o'}\right) \sigma' \frac{h}{h_o} \exp \mu(H_o - H) \tag{2.17}$$

If the back tension is increased then this has the effect of displacing the neutral plane towards the exit of the roll gap. If it becomes large enough then the neutral plane may lie outside the roll gap. Under these conditions the rolls are moving faster than the stock at all positions in the roll gap and slipping of the rolls occurs.

Example -

(a) Assuming no roll flattening and the data for single stand rolling of aluminium set out in Table 2.1, determine the new value of rolling load with an applied front (outgoing) tension of 25 N/mm^2 and an applied back (ingoing) tension of 15 N/mm^2.

(b) Determine also the magnitude of the back tension required to produce slipping between the rolls and the stock in the absence of front tension.

(a) The normal roll pressures in the absence of tension are multiplied by:

for back tension $\left(1 - \frac{t_o}{\sigma_o'}\right) = 1 - \frac{15}{140} = 0.893$

and for front tension $\left(1 - \frac{t_F}{\sigma_F'}\right) = 1 - \frac{25}{147.7} = 0.831$

The values of normal roll pressure are given in the Table 2.3 (taken from Columns 8 and 11 in Table 2.1).

Table 2.3

θ(rad)	p_x	$p_x\left(1-\frac{t_F}{\sigma_F'}\right)$	p_e	$p_e\left(1-\frac{t_o}{\sigma_o'}\right)$	$p_e\left(1-\frac{31.85}{40}\right)$
0	147.7	122.7	191.2	170.7	147.7
.005	154.8	128.6	183.0	163.4	141.4
.010	162.6	135.1	175.6	156.8	135.7
.015	171.3	142.4	169.1	151.0	130.6
.020	180.7	150.2	163.2	145.7	126.1
.025	191.0	158.7	158.1	141.2	122.1
.030	202.3	168.1	153.6	137.2	118.7
.035	216.0	179.7	149.8	133.8	115.7
.040	227.7	189.2	146.4	130.7	113.1
.045	241.9	201.0	143.6	128.2	110.9
.050	257.0	213.6	141.2	126.1	109.1
.0529	266.3	221.3	140.0	125.0	108.2

It should be noted that the values of flow stress used in finding p_x and p_e refer to the appropriate position in the roll gap, whereas the flow stress in the bracket refers to the flow stress at either the entry or the exit. The data are plotted in figure (2.4).

The area under friction hill is determined as previously

$$\text{Area} = 7.064 \text{ rad N/mm}^2$$

and $\text{Load} = 250 \times 7.064 \times 1500$ see equation (2.4)

$$\text{Load} = \underline{\underline{2.65 \times 10^6 \text{ N}}} = \underline{\underline{270 \text{ tonne}}}$$

i.e. there is a $\left(\frac{3.03 - 2.65}{3.03}\right) \times 100 = 12.5\%$ reduction in rolling load.

(b) In order to move the neutral plane to the exit, the pressure curve with back tension, equation (2.17) must be lowered until it intersects the $\theta = 0$ ordinate at $p = 147.7$ N/mm², figure 2.4.

Without tension it intersects the ordinate at 191.2 N/mm²

$$\text{Thus } 191.2\left[1 - \frac{t_o}{140}\right] = 147.7$$

$$\underline{\text{or}} \quad t_o = 140\left\{1 - \frac{147.7}{191.2}\right\} = 31.85 \text{ N/mm}^2$$

The tension at the onset of slipping is therefore 31.9 N/mm².

The pressure curve, equation (2.17), is plotted in figure 2.4 and the individual values tabulated in Table 2.3.

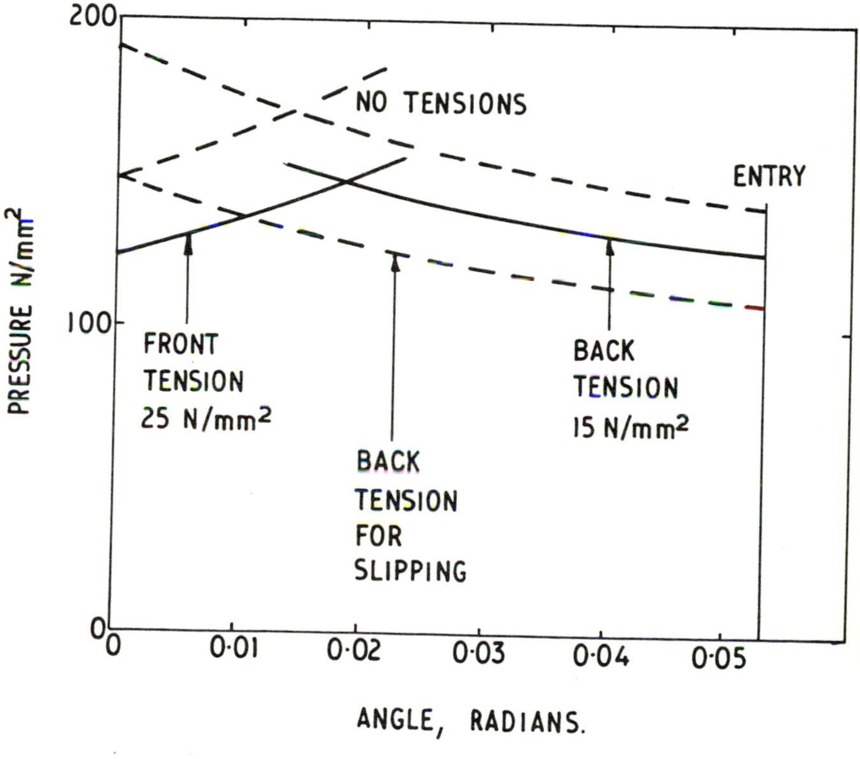
200
100
0
PRESSURE N/mm^2
NO TENSIONS
ENTRY
FRONT TENSION 25 N/mm^2
BACK TENSION FOR SLIPPING
BACK TENSION 15 N/mm^2
0
0·01
0·02
0·03
0·04
0·05
ANGLE, RADIANS.

Fig. 2.4

2.5 Torque Required for Rolling

(a) The most rigorous approach to the determination of torque is by calculating the turning moment applied to the roll barrel necessary to produce the frictional force experienced at the roll surface by the stock.

Considering *one* roll only:

$$\frac{\Gamma}{w} = R\int_{\theta_n}^{\theta_o} \mu p_e \; R'd\theta \; - \; R\int_{o}^{\theta_n} \mu p_x \; R'd\theta \qquad (2.18a)$$

$$= \mu RR'\left\{\int_{\theta_n}^{\theta_o} p_e \; d\theta \; - \; \int_{o}^{\theta_n} p_x \; d\theta\right\}$$

$$\frac{\Gamma}{w} = \mu RR'\left\{\begin{matrix}\text{Area under}\\ \text{entry curve}\end{matrix} - \begin{matrix}\text{Area under}\\ \text{exit curve}\end{matrix}\right\} \qquad (2.18b)$$

The areas are based on the deformed roll radius R' since the frictional forces act over an arc of contact $\theta R'$. Also the *undeformed* roll radius appears in the equation because this is the best value of the length of the turning moment lever arm. It should also be noticed that the frictional force produced by the rolls on the stock at the exit is negative, and this therefore reduces the required torque.

(b) An alternative approach that is more empirical in nature, and therefore liable to greater error, is to assume that the total load acts downwards at a single plane within the arc of contact distant 'a' from the line of roll centres.

Under these conditions the applied torque required to overcome the true roll separating force $F_{R'}$ is given by

$\Gamma = F_{R'}.a$ for *one* roll only (2.19)

The problem lies in finding an accurate value of 'a'. If λ is the ratio of the lever arm 'a' to the projected length of the (undeformed) arc of contact then

$$a = \lambda\sqrt{R\Delta h} \qquad (2.20)$$

This equation cannot be used directly since the rolls deform under load to radius R' and the exit position of the stock is no longer along the line between roll centres. Under these circumstances another parameter similar to λ may be defined, λ', that is related to λ by (Ford 1948)

$$\lambda = \frac{1}{2}\sqrt{\frac{R}{R'}} - \left(\frac{1}{2} - \lambda'\right)\sqrt{\frac{R'}{R}} \qquad (2.21)$$

If a value of λ' is known, and the deformed roll radius R' can be determined from Hitchcock's formula, then λ is established.

The value of load that is needed in the torque equation is the *total* rolling load (even though some of this may act outside the projected arc of contact of the undeformed rolls).

The effect of the surface quality of the rolls on the value of λ' has been examined experimentally (Ford 1948) and two empirical values have been proposed to cover most rolling conditions. For the early passes in cold rolling where the rolls are rough and high frictional conditions exist than the best value is given by

$$\lambda' \simeq 0.43$$

For rolling in the finishing stands where the rolls are highly polished and the friction is low then

$$\lambda' \simeq 0.48$$

It is important to note that they are only average values taken over a range of experimental conditions. The value in any particular pass may differ significantly from these average values. In fact Larke (1967) quotes a whole series of values in tabular form. The standard deviation of these data given in his Table LII is $\sigma_{n-1} = 0.042$ for mean 0.428.

Since rolling mill rolls are usually driven by the same motor the torque required by the upper and lower roll is twice the value calculated above.

Example - Using the data in section 2.3 find the torque required on each roll to make the necessary rolling reduction using both methods of determination.

The data from previously are

$\mu = 0.06 \quad R = 250 \text{ mm} \quad R' = 265.4 \text{ mm} \quad w = 1500 \text{ mm}$

(a) From figure 2.3 the deformed area under the *entry* curve is 5.643 N/mm^2 (x radian) or under the *exit* curve is 2.245 N/mm^2 (x radian)

Hence the torque is, from equation (2.18b)

$$\Gamma = 1500 \times 0.06 \times 250 \times 265.4 \{5.643 - 2.245\}$$

$$= 2.029 \times 10^7 \text{ mmN}$$

$$\Gamma = 2.04 \times 10^4 \text{ mN}$$

(b) Using the second approach, the surface quality of the rolls can be assumed to be smooth since μ is very small. Hence the required value of $\lambda' = 0.48$.

Substituting for λ in equation (2.21)

$$\lambda = \frac{1}{2}\sqrt{\frac{250}{265.4}} - (0.5 - 0.48)\sqrt{\frac{265.4}{250}}$$

$$= 0.4853 - 0.0206 = 0.4647$$

In equation (2.20)

$$a = 0.4647\sqrt{250 \times 0.7}$$

$$= 6.147 \text{ mm}$$

The corrected value of roll separating force $F_{R'}$ has already been found in section 2.3

$$F = F_{R'} = 3.185 \times 10^6 \text{ N}$$

and so in equation (2.19)

$$\Gamma = 6.147 \times 3.18 \times 10^6$$

$$\Gamma = 1.95 \times 10^7 \text{ mmN}$$

$$\Gamma = 1.95 \times 10^4 \text{ mN}$$

The correspondence between the two results is very satisfactory although perhaps somewhat fortuitous.

Had we chosen $\lambda' = 0.43$:

$$\lambda = 0.4853 - (0.5 - 0.43)\sqrt{\frac{265.4}{250}}$$

$$= 0.4853 - 0.072$$

$$= 0.413$$

and $a = 5.464$ mm

Hence $\Gamma = 1.74 \times 10^4$ mN

This figure is significantly smaller than the previous one; and therefore the subjective assessment as to whether the rolls are "rough" or "smooth" can make an important difference to the calculated value of torque.

2.6 Power Rating of the Mill

As the rolls rotate, both the upper and lower rolls do work against the roll separating force. In the previous section the load acted downwards effectively at a position 'a' from the line between roll centres. In one revolution the line of application of the load thus moves through a distance of $2\pi a$.

Hence with rolls operating at N revolutions per minute

Total work per minute for *two* rolls

$$\omega_R = 2F \cdot 2\pi a \cdot N \qquad (2.22a)$$

$$= 4\pi \Gamma N \text{ J/min} \qquad (2.22b)$$

Thus dividing by 60 will give the number of watts of energy developed by the motor in order to make the reduction. Since many motors are still rated in terms of the Imperial Unit "Horse Power" and one h.p. is ~ 746 J/s (W) then dividing by 746 will yield the horse power required to overcome the roll separating force.

The roll neck bearings similarly resist the turning motion and thus extra energy is expended.

If the necks have diameter d and bearing friction is μ_B, then since there are 4 bearings for *two* rolls, the total work done ω_B per minute

$$\omega_B = 4 \left(\mu_B \frac{F}{2}\right) \cdot \left(2\pi \frac{d}{2}\right) N$$

$$= 2 \mu_B F \pi dN \text{ J/min} \qquad (2.23)$$

The overall power required for rolling is therefore the sum $(\omega_R + \omega_B)$ with a small additional amount to account for electrical and transmission losses. On four high mills the work roll bearings do not have to support the roll load as this is done by the backing roll bearings (Roberts 1978) and under these conditions ω_B must refer to the work done by the backing rolls.

The frictional coefficient in the bearings can vary widely and on bronze type grease bearings can be as high as 0.1. However, modern bearings of the roller or oil film type have coefficients as low as 0.001 or 0.002. This makes a very significant difference to the required power. (Larke 1967).

The total amount of electricity consumed in making a particular reduction is obtained from the rate of working given in equations (2.22 and 2.23) and the time taken to roll the material.

Thus the number of kWh consumed

$$= \text{(Work done/min)(Time of Rolling in mins)} \div 3.6 \times 10^6 \qquad (2.24)$$

since 1 kWh = 3.6×10^6 J.

<u>Example</u> - A coil of brass, 35 m long by 0.7 m wide, is produced after a reduction from 5 mm down to 4 mm in a single pass on a two high mill under constant load conditions. The steel roll diameters are 0.6 m, the diameters of the roll neck bearings are 0.5 m and the rolls rotate at 30 r.p.m. If the constant load is 600 tonne, determine the horse power developed, the energy consumed and the percentage of the total work expended in overcoming bearing friction for:

(a) roller bearings $\mu = 0.002$

(b) bronze bearings $\mu = 0.06$

The rolls *have a matt finish ground on them.*

To calculate the rolling time, assume for simplicity that the peripheral speed of the rolls and the speed of the exiting stock are the same.

At 30 rpm,

Linear speed of stock is $\dfrac{30 \times 2\pi}{60}\left\{\dfrac{0.6}{2}\right\}$ m/s

$= 0.943$ m/s

Hence to roll 35 m of strip takes

$$t = \frac{35}{0.943} = 37.1 \text{ s.}$$

A value of 'a' is needed for substitution into the equations. Thus λ must first be found from equation (2.20) using $R = 0.3$ m, $\lambda' = 0.43$, and R' calculated from Hitchcock's formula (section 2.3).

$$R' = R\left\{1 + \frac{CF_{R'}}{\omega\Delta h}\right\} \quad \text{where } C = 2.13 \times 10^{-11}\ m^2/N$$

for steel rolls

$$R' = \frac{0.6}{2}\left\{1 + \frac{2.13 \times 10^{-11} \times 600 \times 10^3 \times 9.81}{0.7 \times 1 \times 10^{-3}}\right\}$$

$$R' = \frac{0.6}{2}(1 + 0.179) = \frac{0.708}{2} = 0.354 \text{ m}$$

From equation (2.21)

$$\lambda = 0.5\sqrt{\frac{R}{R'}} - (0.5 - \lambda')\sqrt{\frac{R'}{R}}$$

$$= 0.5\sqrt{\frac{0.6/2}{0.708/2}} - (0.5 - 0.43)\sqrt{\frac{0.708/2}{0.6/2}}$$

$$\lambda = 0.384$$

Substituting in equation (2.20)

$$a = \lambda\sqrt{R\Delta h}$$

$$a = 0.384\sqrt{\frac{0.6}{2} \times 1 \times 10^{-3}}$$

$$\underline{a = 6.65 \times 10^{-3} \text{ m}}$$

From equation (2.22)

$$\omega_R = 4\pi\Gamma N \text{ J/min}$$

and Γ refers to torque calculated with respect to the *deformed* roll radius R'. If the load is calculated then

$\omega_R = 4\pi F_{R'} aN$ and the true value of load is used.

$$= 4\pi\ 600 \times 10^3 \times 9.81 \times 6.65 \times 10^{-3} \times 30$$

$$= 1.476 \times 10^7 \text{ J/min}$$

From equation (2.23)

(a) low bearing friction :

$$\omega_B = 2\ \mu_B\ F_R,\ \pi\ dN\ J/min$$

$$= 2 \times 0.002 \times 600 \times 10^3 \times 9.81 \times \pi \times 0.5 \times 30$$

$$\omega_B = 0.111 \times 10^7\ J/min$$

(b) high bearing friction

$$\omega_B = 3.329 \times 10^7\ J/min$$

The h.p. developed is therefore

$$h.p. = \frac{\omega_R + \omega_B}{60 \times 746}$$

$= \underline{355}$ - for low friction conditions

and $\underline{1070}$ - for high friction conditions

The energy consumed in making the reduction, from equation (2.24) is

$$\frac{(\omega_R + \omega_B) \times t}{60 \times 3.6 \times 10^6}$$

(a) low bearing friction $= \frac{1.587 \times 10^7}{60} \times \frac{37.1}{3.6 \times 10^6} = \underline{2.73\ kWh}$

(b) high bearing friction $= \frac{4.805 \times 10^7}{60} \times \frac{37.1}{3.6 \times 10^6} = \underline{8.26\ kWh}$

The percent of work done in overcoming bearing friction is

$$\frac{\omega_B}{\omega_R + \omega_B} \times 100$$

Hence

(a) low bearing friction

$$\% \text{ of total} = \frac{0.111 \times 10^7}{1.587 \times 10^7} \times 100 = 7.0\%$$

(b) high bearing friction

$$\% \text{ of total} = \frac{3.329 \times 10^7}{4.805 \times 10^7} \times 100 = 69.3\%$$

It should be noted that with high bearing friction more energy is consumed in overcoming the resistance in the bearings than is consumed in making the rolling reduction.

2.7 Effect of Elastic Extension of the Mill

When material passes through the roll gap the roll separating force extends the size of the "static" roll gap due to the elastic extension of the mill housing, the chocks, screws etc. As the load increases so the mill extends in a linear manner (except for a small non-linear portion at the beginning where the bearings etc. are settling in). The roll gap therefore also expands in a linear manner with load. The gradient of the load-extension curve is called the mill modulus, M. The size of the roll gap h_1 during rolling with load F_1 is given by

$$h_1 = s_o + \frac{F_1}{M} \qquad (2.25)$$

where s_o is the size of the "static" roll gap before rolling. This is clearly the size of the material on exit from the mill.

When the metal passes through the roll gap having an initial thickness h_o then for any given reduction to thickness h_f there will be an associated rolling load that will be larger the larger is the size of the reduction. Thus it is possible to plot a curve of load F against final thickness h_f for a fixed ingoing thickness h_o. The steepness of the curve will depend on the geometry of rolling and the work hardening of the material. This curve is called the "plastic" curve of the material.

These two curves can be combined on a single graph of load against final material thickness. The point of intersection of the graphs represents the exit thickness for an entry thickness h_o, with "static" roll gap s_o and with an associated rolling load F_1, figure (2.5).

If the incoming material increases in thickness then the plastic curve will be displaced to the right giving a new exit thickness h_2 associated with load F_2. The roll gap setting s_o would consequently have to close to position s_1 in order to ensure that the exit thickness of metal were restored to h_1. At this position the new value of rolling load would be F_3.

This diagram represents the basis on which the gauge of metal strip is controlled. The load changes are monitored so that at any time the precise value of the strip thickness is known and hence remedial action in the form of altering the roll gap setting (or changing strip tension) can be accomplished. This type of control is known as "Gaugemeter" control. The refinement of this control technique known as "Feed-forward" control, where the incoming gauge error is monitored, still uses the same principle as above for determining the magnitude of the remedial action (Wright 1978).

<u>Example</u> - Steel strip, 90 mm wide, is cold rolled on an experimental two high mill. The modulus of the mill is 1.4×10^6 N/mm extension, and the work rolls are 125 mm radius. The steel work hardens according to

$$\sigma = K\varepsilon^n$$

where n = 0.4 and K = 1300 N/mm^2

The rolling load, F, is given to a first approximation by equation (2. 7) as

$$F = 1.1\ \bar{\sigma}'\ w\sqrt{R\Delta h}$$

where $\sqrt{R\Delta h}$ is the length of the arc of contact

w is the width

and $\bar{\sigma}'$ is the average plane strain flow stress.

(a) If the unloaded roll gap setting is 0.7 mm and the incoming material is 1.5 mm thick, determine graphically the exit thickness and the instantaneous value of the rolling load.

(b) If the incoming gauge increases to 1.6 mm, determine by how much the roll gap must close to maintain the same exit thickness, and the simultaneous value of the rolling load.

Since the strip is rolled under plane strain conditions we can replace σ and ε in the above equations and obtain

(See equations 1.21 and 1.22):

$$\sigma' = K\left(\frac{2}{\sqrt{3}}\right)^{n+1}(\varepsilon')^n$$

$$\sigma' = 1300(1.155)^{1.4}(\varepsilon')^{0.4}$$

Also for any strain ε' integration gives:

$$\bar{\sigma}' = K\left(\frac{2}{\sqrt{3}}\right)^{n+1}\frac{(\varepsilon')^n}{n+1}$$

$$\bar{\sigma}' = 1136(\varepsilon')^{0.4}$$

Two lines must be constructed on a graph of rolling load against thickness. The first represents the mill modulus, passes through the 0.7 mm point on the axis and has a gradient of 1.4×10^6 N/mm. The second line is curved and is calculated from the load equation for different reductions Δh and for different values of the average flow stress $\bar{\sigma}'$. Values are shown in Table 2.4 and plotted in figure 2.6.

Table 2.4

	h_1, Δh_1	h_2, Δh_2	h_3, Δh_3	h_4, Δh_4	h_5, Δh_5
	1.4, 0.1	1.25, 0.25	1.15, 0.35	1.0, 0.5	0.75, 0.75
ε'	-0.069	-0.182	-0.266	-0.406	-0.693
$\bar{\sigma}'$	390	575	669	792	981
$\sqrt{R\Delta h}$	3.54	5.59	6.61	7.91	9.68
Load, N x 10^5	1.36	3.18	4.38	6.20	9.40

(a) The two curves intersect at 1.075 mm, so this is the outgoing thickness.
The load can be read off as 5.25×10^5 N

(b) A new plastic curve must be constructed parallel to the first and intersecting the axis at 1.6 mm. If the roll gap were not altered then the new rolling load would be 5.9×10^5 N.

As the roll gap closes the modulus line is moved to the left until it intersects the new plastic curve at the position of the original outgoing thickness, i.e. 1.075 mm. The intersection on the axis now gives the unloaded roll gap necessary to give this gauge of material. This is seen to be 0.61 mm and the corresponding rolling load is 6.45×10^5N. The roll gap must therefore close by 0.09 mm.

It is coincidence that this diagram is symmetrical. It results from the value of the modulus of mill chosen. In production, the mills have a much lower value of modulus and in order to obtain large reductions the roll gap would have to close to zero (and be loaded if compressive prestress is possible) so that a realistic reduction could be achieved. This has led to the development of prestressed mills and modification of older mills, the design of which effectively produces a mill with a very high mill modulus.

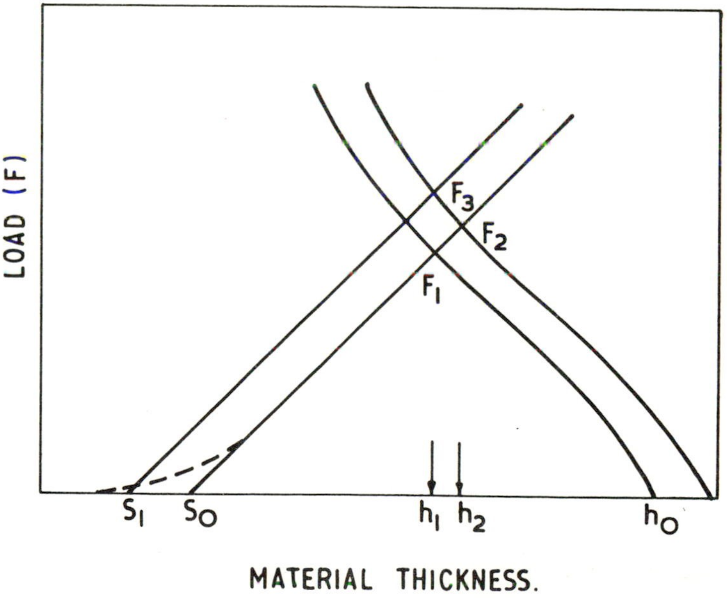

Fig. 2.5

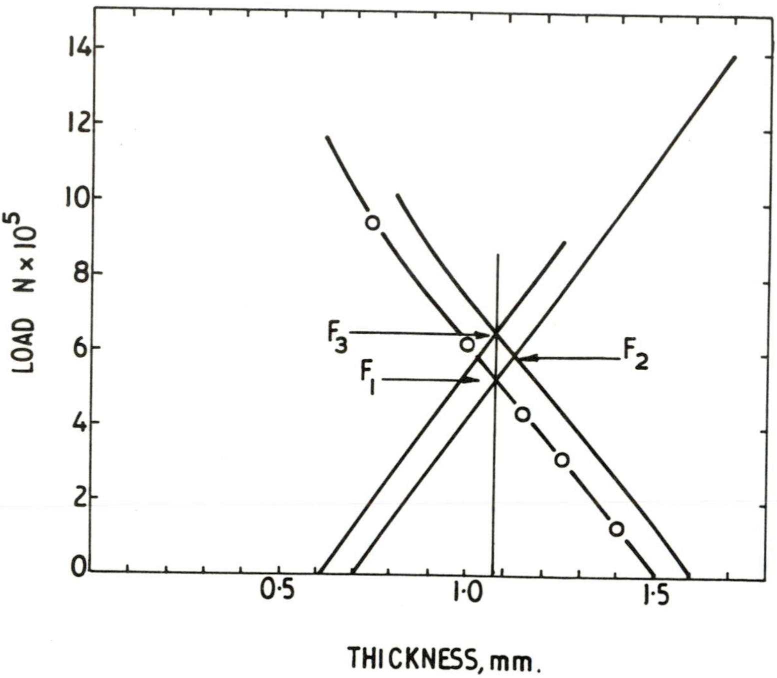

Fig. 2.6

2.8 Roll Camber

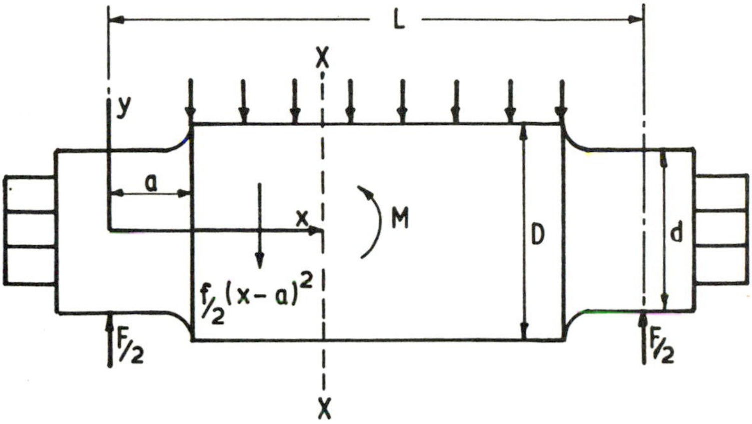

The amount that the roll barrel deflects under load can be calculated from elementary beam theory for simple bending. This enables a crown, i.e. a camber, to be ground on the roll barrel that just counterbalances the deformation of the roll.

If the roll separating force is F then the roll neck bearings generate an upward force $\frac{F}{2}$ effectively at some position 'a' away from the roll neck junction (see diagram). The position of 'a' can be estimated from a knowledge of the "design and operating experience of the mill" (Larke 1967).

For simplicity of illustration consider the strip extending across the whole length of the roll barrel. This simplifies the equations but the principle remains the same. The weight of material in the roll may also be neglected since it is small in relation to the rolling loads. If the rolls are very large then the load can also be expected to be very large. If the rolls are only small then the rolling load will swamp the effect of the weight of the roll material. (The weight of the rolls will tend to increase the deflection of the lower roll, and decrease the deflection of the upper).

Referring to the above figure, the bending equation at some plane XX distance x from the zero of the axes is

$$EI_D \frac{d^2 y_B}{dx^2} = \frac{F.x}{2} - \frac{(x - a)^2}{2} f$$

using the convention that a clockwise moment is positive.

D is the roll diameter, E is Young's Modulus,

I_D is the moment of inertia of the roll $\left[= \frac{\pi}{4} \left(\frac{D}{2}\right)^4 \right]$

and f is the distributed force per unit width acting across the roll barrel, $\left[= \frac{F}{L-2a} \right]$.

This equation can be integrated twice giving

$$2EI_D \, y_B = \frac{Fx^3}{6} - \frac{f}{12}(x - a)^4 + C_1 x + C_2 \qquad (2.27)$$

where C_1 and C_2 are the constants of integration given by

$$C_1 = \frac{f}{3}\left(\frac{L}{2} - a\right)^3 - \frac{FL^2}{8}$$

and $$C_2 = -\frac{Fa^3}{6} + \frac{FL^2}{8} a - \frac{fa}{3}\left(\frac{L}{2} - a\right)^3$$

The deflection due to shear is obtained in a similar way. The slope of the deflection curve is approximately equal to the shear strain at the neutral axis and therefore (Timoshenko and Gere 1973):

$$\frac{dy_S}{dx} = \frac{\alpha F}{GA}\left\{\frac{x - a}{L - 2a} - \frac{1}{2}\right\} \qquad (2.28)$$

where G is the shear modulus and A is the area of cross section of the roll barrel. (The shear force is positive when the action of the right hand side of XX *on* the left hand side of XX tends to shear it in the positive y-direction).

α is a constant that for circular sections has a value of $\frac{4}{3}$.

Integrating (2.28) and putting in the boundary conditions i.e. when x = a or (L - a), y = 0 gives:

$$y_s = \frac{\alpha F}{GA}\left\{\frac{(x-a)^2}{2(L-2a)} - \frac{1}{2}(x-a)\right\} \qquad (2.29)$$

The total deflection under load can therefore be obtained by adding the deflections due to both bending and shear

$$y_{Total} = y_B + y_s \qquad (2.30)$$

Having obtained the total deflection a crown is then ground on the rolls in such a manner that when under load the cambered rolls will deflect back to a flat profile. Thus the curvature of the crown is the exact mirror image of the curvature produced on the rolls by the deflection.

The camber is associated with a fixed rolling load and a fixed width of strip. There are, however, techniques that enable narrower strip to be rolled successfully using the same rolls (Larke 1967).

In addition to considerations of roll bending, care must also be taken to ensure that the rolling loads generated do not lead to fatigue cracking at the roll necks or on the roll barrel surface.

<u>Example</u> - Calculate the deflection curve due to bending and shear, and hence the required camber, for cold rolling at a constant rolling load of 6 MN on a two high mill with work rolls 0.5 m diameter, if the strip is 1.2 m wide and extends across the whole width of the barrel. The distance from the line of action of the load on the bearing to the roll neck is 0.15 m. It may be assumed that Young's modulus is 216 GN/m^2 and the shear modulus is 82 GN/m^2.

From the known density of steel and the roll dimensions the weight/unit width of roll barrel due to the rolls is only about 0.3% of the weight/unit width due to the roll separating force and therefore can be neglected.

$L - 2a = 1.2$ m, $\therefore$ $L = 1.5$ m

Also,

$$f = \frac{F}{L - 2a} = \frac{6 \times 10^6}{1.2} = 5 \times 10^6 \text{ N/m}$$

In equation (2.27) the Moment of Inertia of the roll barrel about the axis of bending is given by: (Timoshenko and Gere 1973)

$$I_D = \frac{\pi}{4}(0.25)^4 = 3.068 \times 10^{-3} \text{ m}^4$$

The constants of integration C_1 and C_2 are thus given by

$$C_1 = \frac{5 \times 10^6}{3}(0.6)^3 - \frac{6 \times 10^6(1.5)^2}{8} = -1.328 \times 10^6$$

$$C_2 = -\frac{6 \times 10^6 \times (0.15)^3}{6} + \frac{6 \times 10^6(1.5)^2(0.15)}{8} - \frac{5 \times 10^6 \times 0.15(0.6)^3}{3} = 0.1958 \times 10^6$$

<u>When x = 0.3 m</u> :

In bending: equation (2.27)

$$y_B = \left\{\frac{1}{2 \times 216 \times 10^9 \times 3.068 \times 10^{-3}}\right\} \left\{\frac{6 \times 10^6}{6}(0.3)^3 - \frac{5 \times 10^6}{12}(0.15)^4 - 1.328 \times 10^6 \times 0.3 + 0.1958 \times 10^6\right\}$$

$$\therefore \quad y_B = -0.1326 \text{ mm}$$

In shear: equation (2.29)

$$y_S = \frac{1.33 \times 6 \times 10^6}{82 \times 10^9 \times \pi(0.25)^2} \left\{\frac{(0.15)^2}{2(1.2)} - \frac{1}{2}(0.15)\right\}$$

$$y_S = -0.0326 \text{ mm}$$

The total deflection is therefore:

$$y_T = -0.1326 - 0.0326 = -\underline{\underline{0.1652 \text{ mm}}}$$

The values for different values x are laid out in Table 2.5 and are plotted in figure (2.7).

Table 2.5

x	0.15	0.3	0.45	0.6	0.75	0.90	1.05	1.20	1.35
$-y_B$	0	0.1326	0.2370	0.3033	0.3262	0.3033	0.2370	0.1326	0
$-y_S$	0	0.0327	0.0560	0.0700	0.0747	0.0700	0.0560	0.0327	0
$-y_T$	0	0.1653	0.293	0.3733	0.4009	0.3733	0.2930	0.1653	0

It is evident from the curves that although the shear contribution to the deflection is much smaller than the bending contribution, it cannot be neglected. In addition to the deflection due to load there is a small effect of roll heating on the shape of the crown. However, its effect is only of secondary importance in cold rolling, and can be neglected; in hot rolling, on the other hand, it can dominate the effects of roll bending.

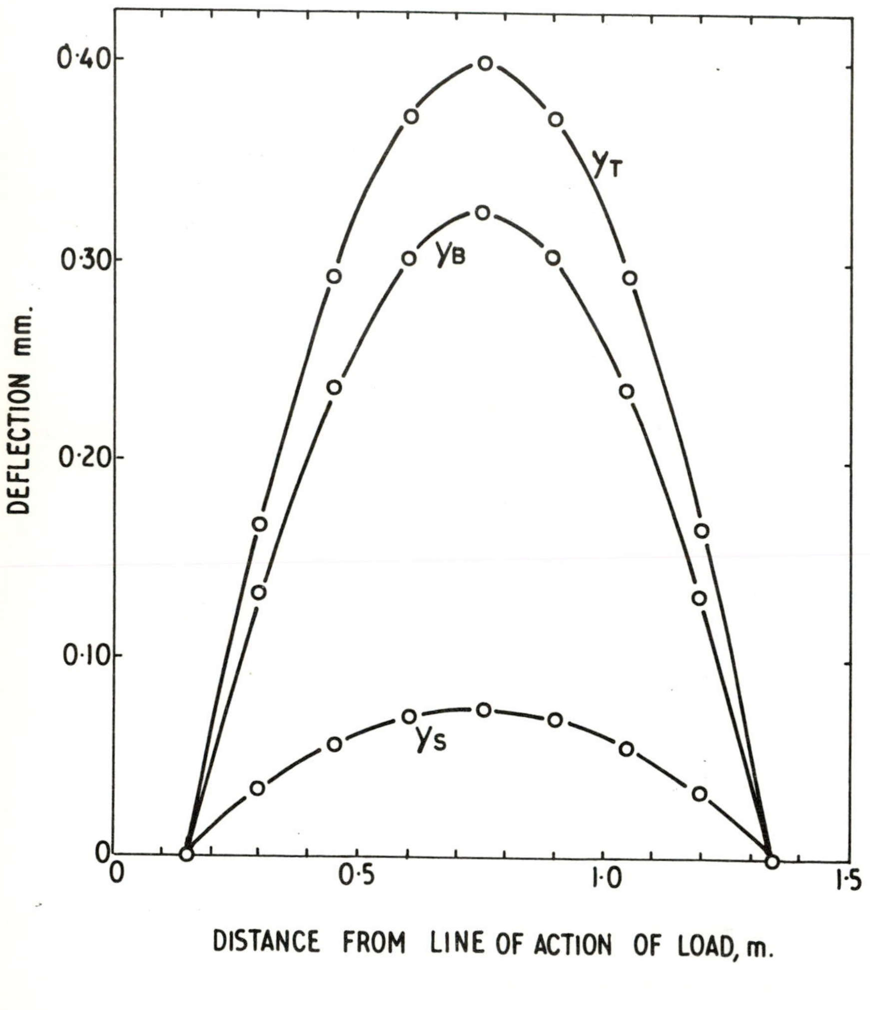
0·40
0·30
0·20
0·10
0
DEFLECTION mm.
y_T
y_B
y_S
0
0·5
1·0
1·5
DISTANCE FROM LINE OF ACTION OF LOAD, m.

Fig. 2.7

SECTION 3 - HOT ROLLING

When metals are hot rolled, it is useful to be able to calculate the loading which the deformation will impose on the mill, so as to ensure that the mill is not overloaded and that there is sufficient power available for the reduction to be made.

The rolling load will depend on the geometry of the reduction, i.e. the initial width and thickness of the stock, the work roll diameter and the degree of the reduction, on the frictional conditions existing at the roll-stock interface and on the flow stress of the metal being rolled.

As has already been mentioned in section 1, the flow stress of a metal at high temperatures is dependent upon its composition, its microstructure and on the strain, strain rate, temperature and deformation mode, which for rolling implies the extent to which the reduction deviates from plane strain.

3.1 The Strain Rate in Hot Strip Rolling

In a roll pass the strain rate varies from a maximum value immediately after entry into the roll gap to zero at the exit. The form of this variation can be shown graphically in the following manner. At any point in the roll pass the instantaneous value of plane strain rate is given by:

$$\dot{\varepsilon}' = \frac{dh}{dt} \cdot \frac{1}{h} \qquad (3.1)$$

where h is the thickness of the strip at that point.

Now, as seen from the geometry of rolling,

$$\frac{dh}{dt} = 2\,v\,\sin\theta \qquad (3.2)$$

where v is the peripheral speed of the work roll (conditions of sticking friction are assumed for hot rolling).

Also

$$h = h_f + 2R(1 - \cos\theta) \qquad (3.3)$$

where R is the work roll radius.

Hence, in equation (3.1)

$$\dot{\varepsilon}' = \frac{2\ v\ \sin\theta}{h_f + 2R(1 - \cos\theta)} \qquad (3.4)$$

Geometry of Rolling

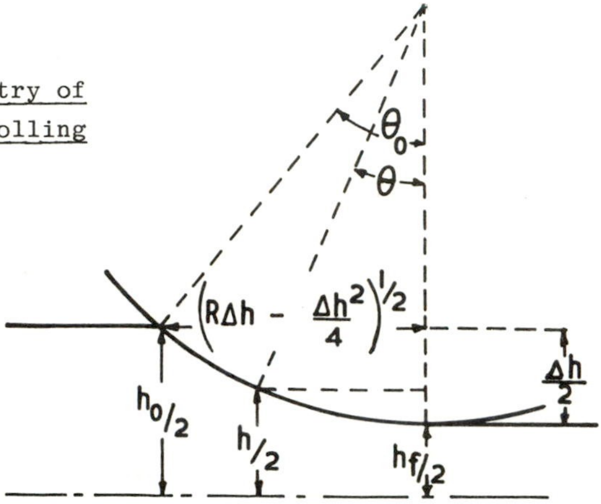

Thus by subdividing the arc of contact into a suitable number of sectors, each subtending an equal angle dθ at the roll centre, a plot of the variation of $\dot{\varepsilon}'$ through the roll gap can be obtained, as shown in the example below.

In order to determine the flow stress of the material in the roll gap it is usual to calculate the mean strain rate for the pass; in terms of time this is given by:

$$\dot{\varepsilon}'_{(t)} = \frac{1}{\theta_0} \int_{\theta_0}^{0} \frac{2\ v\ \sin\theta}{h_f + 2R(1 - \cos\theta)}\ d\theta \qquad (3.5)$$

$$= \frac{v}{R\theta_0}\ \ell n\ \frac{h_f}{h_o} \qquad (3.6)$$

$$\approx -\frac{v}{\sqrt{R\Delta h}}\ \ell n\ \frac{h_o}{h_f} \qquad (3.7)$$

Note that although in practice the negative sign is usually ignored its presence signifies that the strain rate is being calculated for compressive deformation conditions.

<u>Example</u> - Plot the variation of $\dot{\varepsilon}'$ through the roll gap when mild steel strip 50 mm thick, at 1000°C, is reduced in thickness by 36.46% under plane strain conditions in a single pass. The work rolls are 1200 mm in diameter and are revolving at 1 revolution per second.

From the geometry of rolling, the entry angle, θ_o, is given by:

$$\cos\theta_o = \frac{R - \frac{\Delta h}{2}}{R} = \frac{600 - 9.115}{600}$$

therefore $\theta_o = 10^o$

The speed of the stock = the peripheral speed of the rolls

$$\therefore \quad v = 2\pi 600 \times 1 = 3770 \text{ mm s}^{-1}$$

The values of $\dot{\varepsilon}'$ v θ can now be tabulated as follows, using equation (3.4):

<u>Table 3.1</u>

θ deg	10	9	8	7	6	5	4	3	2	1
$\dot{\varepsilon}'$ s^{-1}	26.18	25.35	24.14	22.57	20.56	18.08	15.16	11.81	8.10	4.12

and plotted in figure 3.1, which thus shows the form of the variation of $\dot{\varepsilon}'$ with position in the roll gap. The mean value of $\dot{\bar{\varepsilon}}'$, which is necessary for determining the average flow stress of the metal in the pass, can now be found either from the graph or more rapidly by using equation (3.7). In the latter manner:

$$\dot{\bar{\varepsilon}}'_{(t)} = \frac{3770}{\sqrt{600 \times 18.23}} \ \ell n \frac{50}{31.77}$$

$$= \underline{16.35 \text{ s}^{-1}}$$

This calculated value has been plotted on the graph in

figure 3.1.

The mean strain rate can also be calculated in terms of a strain average, rather than the time average used here, or in terms of a logarithmic relationship (Sellars, 1981). However, although these alternate methods, for this example, produce values of $\bar{\dot{\varepsilon}}'$ greater by 23% and by 13% respectively compared with the above figure, the increase in flow stress which these higher values give is relatively small.

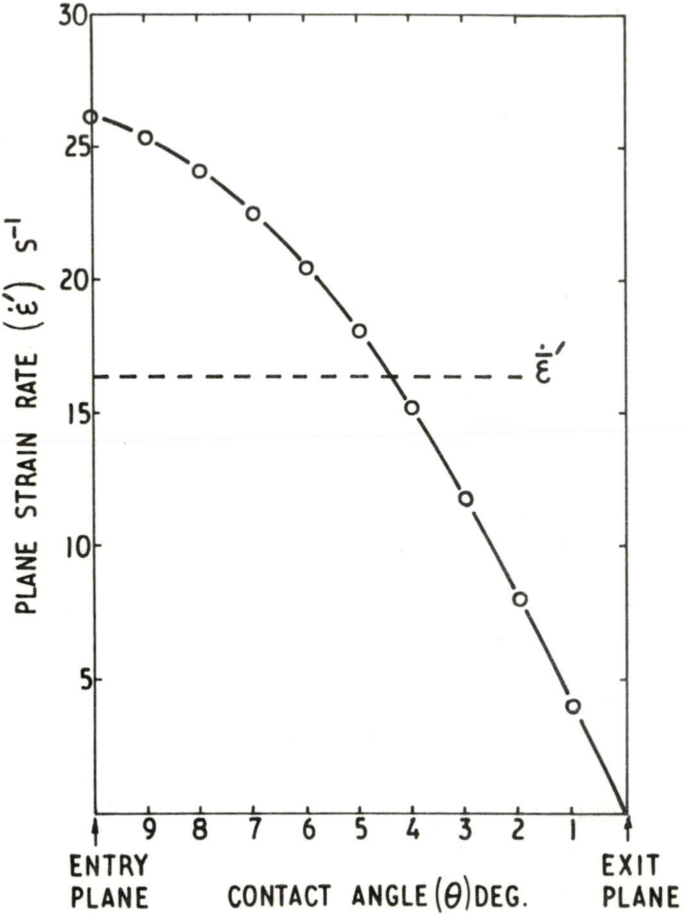

Fig. 3.1

3.2 Hot Rolling Loads

The most widely used expression for calculating the hot rolling load, F, under plane strain deformation conditions is that due to Sims (1954) which states that

$$F = w \, \bar{\sigma}' \sqrt{R \Delta h} \; Q \qquad (3.8)$$

where w is the width of the stock being rolled, $\sqrt{R\Delta h}$ is the projected arc of contact and Q is a complex function of the roll radius, the final thickness of the stock and the reduction in the pass. Values of Q for different deformation geometries given by Larke (1957), are shown in Table 3.2. The required value of $\bar{\sigma}'$, the mean plane strain flow stress, must be computed for the particular material, strain, mean rate of plane strain deformation and temperature pertaining in the roll pass.

Table 3.2 [(After Larke (1957)]

$\frac{R}{h_f}$	Values of Q for per cent Reduction of						
	5	10	20	30	40	50	60
2	0.857	0.880	0.905	0.912	0.904	0.881	0.838
5	0.904	0.948	1.003	1.037	1.055	1.055	1.040
10	0.956	1.022	1.111	1.173	1.216	1.242	1.249
20	1.029	1.127	1.264	1.362	1.439	1.496	1.501
30	1.086	1.208	1.379	1.505	1.608	1.689	1.747
50	1.175	1.337	1.563	1.734	1.875	1.992	2.085
100	1.339	1.570	1.898	2.152	2.365	2.548	2.700
150	1.464	1.698	2.158	2.471	2.740	2.972	3.170
200	1.573	1.901	2.374	2.740	3.056	3.330	3.567
250	1.667	2.036	2.566	2.979	3.334	3.644	3.915
300	1.749	2.157	2.741	3.195	3.586	3.930	4.231

In the example which follows, flow stress data for the material at the required deformation temperature have been obtained from a paper by Cook (1957), figure 3.2. However, these plots are for uniaxial, rather than plane strain, test conditions. Therefore to utilise Cook's data to compute $\bar{\sigma}'$

for use in equation (3.8) the following steps need to be carried out:

(i) Convert the mean strain rate, $\bar{\dot{\varepsilon}}'$, for the roll pass, calculated from equation (3.7) to a mean uniaxial strain rate, $\bar{\dot{\varepsilon}}$, from the relationship:

$$\bar{\dot{\varepsilon}} = \frac{2}{\sqrt{3}} \bar{\dot{\varepsilon}}' \qquad (3.9)$$

(ii) Obtain $\sigma - \varepsilon$ data at this value of $\bar{\dot{\varepsilon}}$ and between the strain limits of the pass. This will usually involve interpolation between data curves as the required value of $\bar{\dot{\varepsilon}}$ is unlikely to be coincident with any of those available. If, at the deformation temperature, the flow stress and strain rate for the material are related through the expression:

$$\dot{\varepsilon} = B \exp \beta \sigma \qquad (3.10)$$

where B and β are constants, then interpolation between these curves must be on a logarithmic scale (see Example 1.5).

The strain limits of the pass are from $\varepsilon_0' = 0$ to $\varepsilon_f' = \ln \frac{h_f}{h_o}$, i.e. from $\varepsilon_o = 0$ to $\varepsilon_f = \frac{2}{\sqrt{3}} \ln \frac{h_f}{h_o}$.

(iii) Calculate $\bar{\sigma}'$ from the $\sigma - \varepsilon$ data. Now, Sims defines $\bar{\sigma}'$ as:

$$\bar{\sigma}' = \frac{1}{\theta_o} \int_o^{\theta_o} \sigma' d\theta \qquad (3.11)$$

where θ is in radians and θ_o is the contact angle, (see figure in section 3.1). Therefore to calculate $\bar{\sigma}'$ for the pass, the $\sigma - \varepsilon$ values must firstly be converted to $\sigma' - \varepsilon'$ values and the ε' values then converted to θ values.

From equation (3.3),

$$h = h_o + 2R(\cos\theta_o - \cos\theta) \qquad (3.12)$$

and as $\varepsilon' = \ln \frac{h}{h_o}$

$$\cos\theta = \cos \theta_o - \frac{h_o}{2R} (\exp \varepsilon' - 1) \qquad (3.13)$$

Use of equation (3.13) and remembering that ε' is negative, even though the negative sign is frequently omitted in reporting values, enables the conversion of ε' to θ values in the pass to be made. The corresponding values of σ' can then be obtained as $\sigma' = 2\sigma/\sqrt{3}$ and $\bar{\sigma}'$, as defined by equation (3.11), can be derived by graphical integration.

Example - By use of Sims equation (3.8) and Cook's data of figure 3.2, calculate the rolling load when a 40 mm thick, 1250 mm wide, low carbon steel plate at $1000^{\circ}C$ is reduced in thickness to 24 mm in one pass, under conditions of plane strain. The steel work rolls are 865 mm diameter and are revolving at 66 revolutions per minute.

Step (i)

From equation (3.7):
$$\dot{\bar{\varepsilon}}' = \frac{2 \times \pi \times 432.5 \times 66}{60\sqrt{432.5 \times 16}} \ell n \frac{40}{24} \ s^{-1}$$
$$= 18.36 \ s^{-1}$$

From equation (3.9):
$$\dot{\bar{\varepsilon}} = \frac{2}{\sqrt{3}} 18.36 = 21.2 \ s^{-1}$$

Step (ii)

In figure 3.2 the $\sigma - \varepsilon$ curve for $\dot{\bar{\varepsilon}} = 21.2$ lies between those of $\dot{\bar{\varepsilon}} = 8$ and $\dot{\bar{\varepsilon}} = 40$. At $1000^{\circ}C$ for this steel the relationship of equation (3.10) holds (Sellars and Tegart, 1972), thus interpolation between these curves must be on a logarithmic scale. Now when $\dot{\bar{\varepsilon}} = 40, 21.2, 8$, $\log_{10} \dot{\bar{\varepsilon}} = 1.6, 1.33, 0.9$.

Thus to obtain the required $\sigma - \varepsilon$ data from figure 3.2, interpolation must be made at $\frac{1.33 - 0.9}{1.6 - 0.9} = 0.61$ of the gap between the $\dot{\bar{\varepsilon}} = 8$ and $\dot{\bar{\varepsilon}} = 40$ curves. The results of interpolation are shown in columns (1) and (2) in Table 3.3. For this pass $\varepsilon' = \ell n \frac{24}{40} = (-)0.5108$, thus $\varepsilon = \frac{2}{\sqrt{3}} 0.5108 = (-)0.59$, which is therefore the strain limit for the required data.

Table 3.3 : σ - ε data for $\dot{\bar{\varepsilon}}$ = 21.2

Column (1)	(2)	(3)	(4)	(5)	(6)
ε *	σ(t.s.i.)	σ' N/mm^2	ε' *	θ radians	e' *
0.05	6.9	123	0.043	0.182	0.042
0.10	8.1	144	0.087	0.171	0.083
0.20	9.6	171	0.173	0.149	0.159
0.30	10.6	189	0.260	0.126	0.229
0.40	11.0	196	0.346	0.100	0.292
0.50	11.3	201	0.433	0.067	0.351
0.59	11.5	205	0.511	0.000	0.400

* negative signs omitted, following common practice.

Step (iii)

Convert σ to σ' by multiplying by $\frac{2}{\sqrt{3}}$ and by 15.44 as the units of σ are in tons.in^{-2}, this gives the results in column (3) of Table 3.3.

The values of ε' in column (4) are obtained by multiplying the ε values by $\frac{\sqrt{3}}{2}$.

Now, $\frac{h_o}{2R} = \frac{40}{865} = 0.0462$ and $h_f = 24$ mm. Thus, from equation (3.3),

$$\cos\theta_o = 1 - \frac{40-24}{865} = 0.9815$$

$$(\theta_o = 11.04^o = 0.193 \text{ radians})$$

Hence, from equation (3.13)

$$\cos\theta = 0.9815 - 0.0462(\exp\varepsilon' - 1)$$

Substituting the values of ε' gives the values of θ shown in column (5).

The values of σ' and θ are plotted in figure 3.3. By graphical integration of this curve a value of $\bar{\sigma}'$ = 183 N/mm^2, as defined by equation (3.11), is obtained.

To calculate the value of Q for the pass, interpolation in Table 3.2 for the reduction from 40 mm down to 24 mm, i.e. a reduction of 40% and $\frac{R}{h_f} = \frac{432.5}{24} = 18.02$, gives Q = <u>1.395</u>.

The rolling load for the pass is then given by equation (3.8) as

$$F = 1250 \times 183 \sqrt{432.5 \times 16} \times 1.395 \text{ N}$$

$$= 26.54 \times 10^6 \text{ N}$$

$$= \underline{26.54 \text{ MN}}$$

Note that in determining $\bar{\sigma}'$ from figure 3.3 the intercept of the curve with the θ_o value is conjectural; thus as drawn the $\bar{\sigma}'$ value may have been over-estimated by inclusion of the shaded area. However, neglecting this area only reduces the $\bar{\sigma}'$ value to 182 N/mm², so the error is negligible.

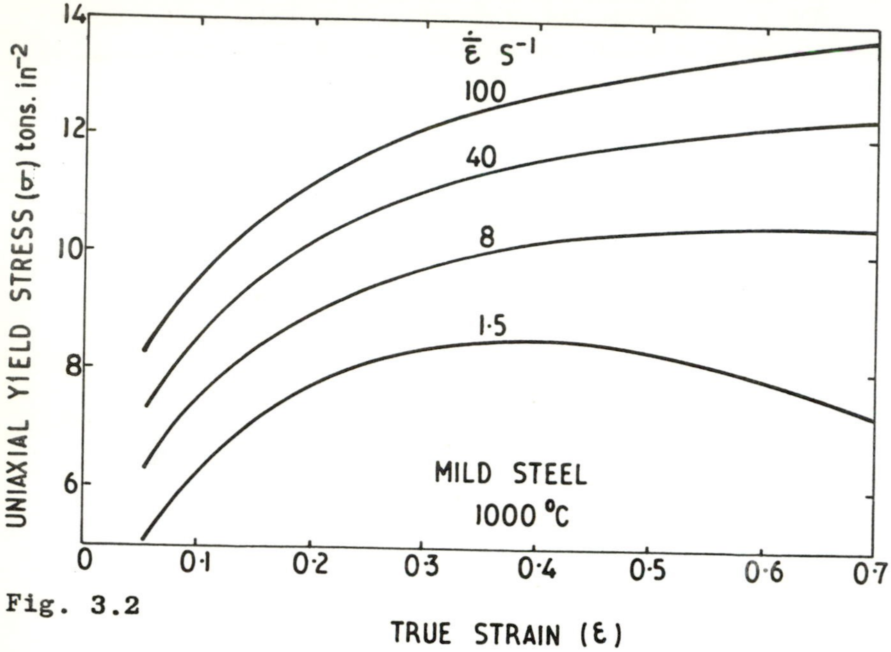
UNIAXIAL YIELD STRESS (σ) tons. in⁻²
ε̇ s⁻¹
100
40
8
1·5
MILD STEEL
1000 °C
14
12
10
8
6
0
0·1
0·2
0·3
0·4
0·5
0·6
0·7
TRUE STRAIN (ε)

Fig. 3.2

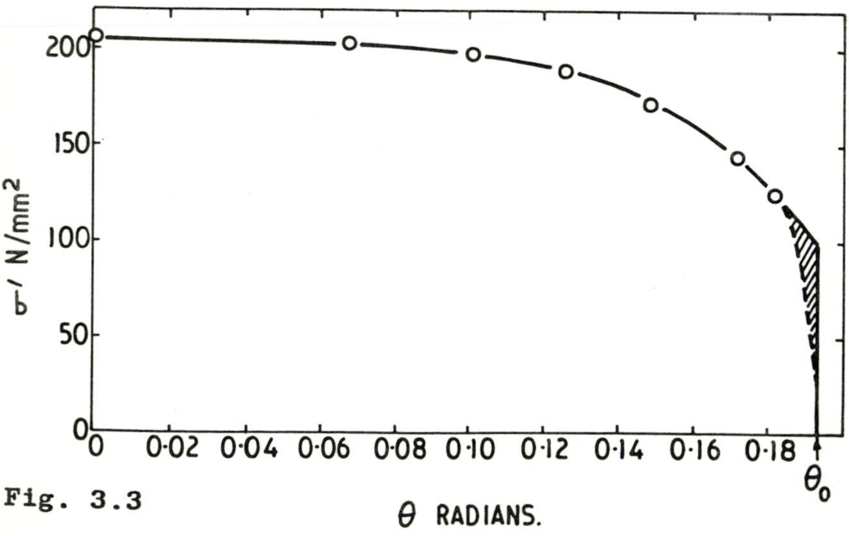
σ' N/mm²
200
150
100
50
0
0
0·02
0·04
0·06
0·08
0·10
0·12
0·14
0·16
0·18
θ₀
θ RADIANS.

Fig. 3.3

3.3 Torque Requirements for Hot Rolling

For the determination of torque a mean plane strain flow stress $\bar{\sigma}'_g$ defined in a different way must be determined. $\bar{\sigma}'_g$ is defined by Sims (1954) as:

$$\bar{\sigma}'_g = \frac{1}{e'_f} \int_0^{e'_f} \sigma' \, de' \qquad (3.14)$$

where e' is the conventional plane strain.

If σ' is only known in terms of ε' it is necessary as a first step to convert the ε' values to e' from thc relationship

$$e' = \frac{h - h_o}{h_o} = \exp \varepsilon' - 1 \qquad (3.15)$$

Thereafter, $\bar{\sigma}'_g$ can then be determined from graphical integration of the σ' v e' curve.

According to Larke (1957) the expression for calculating the torque, Γ, on each roll is:

$$\Gamma = mF \frac{\bar{\sigma}'_g}{\bar{\sigma}'} \sqrt{R.\Delta h}. \qquad (3.16)$$

where $\bar{\sigma}'$ is defined by equation (3.11).

As this expression shows, the lever arm for calculating the torque is the fraction m $\frac{\bar{\sigma}'_g}{\bar{\sigma}'}$ of the projected arc of contact, $\sqrt{R\Delta h}$. The required value of m, which varies with pass reduction and with the ratio $\frac{R}{h_f}$, can be obtained from Table 3.4.

Table 3.4 - Numerical values of m (Larke (1957))

% reduction	$\frac{R}{h_f}$	m
20	10	0.490
30		0.481
40		0.475
50		0.467
20	20	0.487
30		0.480
40		0.475
50		0.467

Note that the method of calculating the lever arm for hot rolling differs from that used for cold rolling. For comparison see section 2.5.

Example - Calculate the roll torque requirements for the rolling conditions of example 3.2.

The values of ε', column (4) in Table 3.3, are converted to the e' values of column (6) in the table using equation (3.15). A plot of σ' v e' is then made, figure 3.4. Graphical integration of this figure between strains of 0.0 and 0.4 gives $\bar{\sigma}'_g$ = 172 N/mm²

For the reduction of 40% and $\frac{R}{h_f}$ value of 18.02, from Table 3.4 m = 0.475, therefore in equation 3.16

$$\Gamma = 0.475 \times 26.54 \times 10^6 \times \frac{172}{183} \sqrt{\frac{432.5 \times 16}{10^6}} \text{ N-m}$$

$$= 9.86 \times 10^5 \text{ N-m}$$

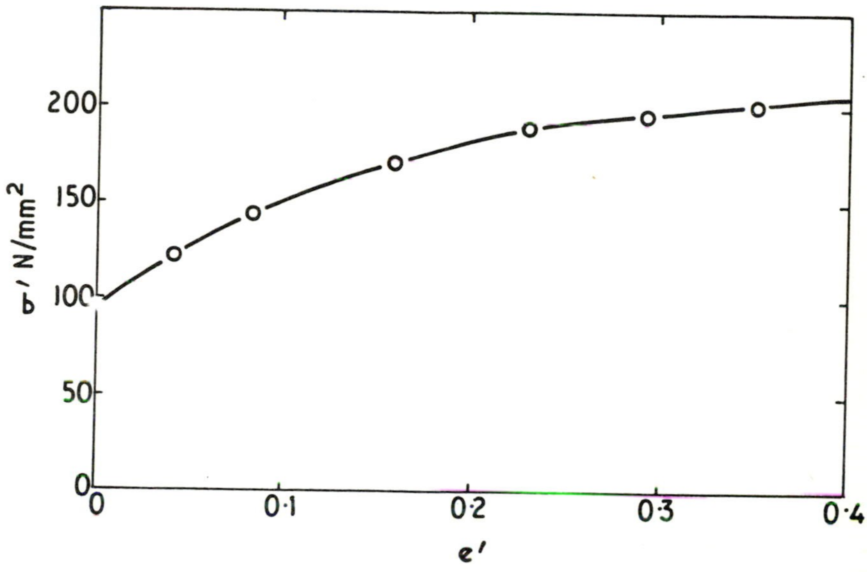

Fig. 3.4

3.4 Power requirements of the mill in hot rolling

A discussion of the energy consumption on primary reversing mills shows that, on average, the reduction of the material takes about 60% of the total amount of energy consumed, and the rest is lost to the bearings, in the motors and generators and in other ancillary equipment (McGannon 1971). To assist in mill design an empirical equation has been proposed that takes these factors into account when assessing torque requirements. It might be argued, therefore, that an accurate assessment of rolling torque is unnecessary. On the contrary, developments in our understanding of hot rolling in recent years suggest that the load and torque may be sensitive indicators of the microstructural changes that occur in the metal during rolling. Therefore a good understanding of the interaction between the material and the rolling parameters is essential.

The power requirements in hot rolling tend to be greater than those in cold rolling due to the much greater thicknesses involved. Thus in hot rolling the energy losses in the bearings are a smaller proportion of the total power requirements than in cold rolling. As for cold rolling, see section 2.6, the power requirements for hot rolling, ω_R, are given by:

$$\omega_R = 4\pi F_R aN = 4\pi \Gamma N \qquad (3.17a)$$

$$\omega_B = 2\pi\mu_B F_R dN \qquad (3.17b)$$

Consideration of these equations reveals that the bearing contribution can be proportionally reduced by reducing the bearing frictional coefficient, and by increasing the arc of contact.

Example - Calculate the power requirement for the rolling conditions of example 3.3, if the roll neck diameter is 0.7 m, μ_B = 0.002 and the rolls operate at 66 rpm.

The roll torque requirements are 9.86×10^5 N-m. Thus from (3.17a).

$$\omega_R = 4\pi\ 9.86 \times 10^5 \times \frac{66}{60}$$

$$\omega_R = \underline{\underline{136 \times 10^5\ W}}$$

$$\omega_B = 2\pi\ 0.002 \times 26.54 \times 10^6 \times 0.70 \times \frac{66}{60}$$

$$= \underline{\underline{2.57 \times 10^5\ W}}$$

It is evident from these results that power lost in the bearings is less than 2% of the total power requirement, and, in view of uncertainties in temperature and flow stresses relating to ω_R, can be ignored.

3.5 Hot rolling load and power requirements allowing for roll flattening

As discussed in section 2.3 the rolls deform elastically under load and therefore correction is needed for the new effective radius of the rolls. Hitchcock's formula may again be used

$$R' = R\left\{1 + \frac{CF_{R'}}{\Delta hw}\right\} \qquad (3.18)$$

The constant C has been defined in equation (2.12); it depends upon the elastic constants of the roll material and typical values for steel rolls used in hot rolling and for cast iron rolls are 2.16×10^{-5} and 4.53×10^{-5} mm^2/N (Cook & McCrum 1958). Values of R, Δh and w are known and a first estimate of the rolling load $F_{R'}$ is obtained by assuming no flattening, i.e. $R' = R$, as in example 3.2.

By substitution of R' for R in equation (3.8) the new rolling load $F_{R'}$ can be calculated. Note that since Q is dependent upon $\frac{R'}{h_f}$ it also has to be recalculated. Further values of $F_{R''}$ and $F_{R'''}$ can be calculated, the process of iteration being continued until it is clear that a sufficiently accurate result has been obtained for the rolling load.

The power requirement corrected for roll flattening is determined by substituting the values of $F_{R'}$ and R' into equation 3.16 and then substituting the corrected value of torque obtained into equation (3.17a).

Example - Calculate the rolling load and power requirements allowing for roll flattening, when using steel rolls, for the rolling conditions of example 3.2.

$F = 26.54$ MN, thus from equation (3.18):

$$R' = 432.5\left(1 + \frac{2.16 \times 10^{-5} \times 26.54 \times 10^{6}}{16 \times 1250}\right)$$

$$= 444.9 \text{ mm}$$

By substitution of R' into equation (3.8) and recalculating

Q when $\frac{R'}{h_f}$ = 18.54 then:

$$F_{R'} = 1250 \times 183 \sqrt{444.9 \times 16} \times 1.406$$
$$= \underline{27.14 \text{ MN}}$$

As in section 2.3, to check if a single recalculation of rolling load is sufficient to allow for roll flattening further values of R'' and R''' can be determined. The results are tabulated below:

Table 3.6

R = 432.5 mm,	F_R = 26.54 MN
R' = 444.9 mm,	$F_{R'}$ = 27.14 MN
R'' = 445.2 mm,	$F_{R''}$ = 27.16 MN
R''' = 445.2 mm,	$F_{R'''}$ = 27.16 MN

These results indicate that in this example one recalculation is sufficient and that when roll flattening is allowed for the rolling load increases by 2.3%.

In the above example $\frac{R'}{R}$ = 1.03; Larke (1957) suggests that if this ratio is $\leqslant$ 1.05 no correction for roll flattening need be made to the roll load, i.e. he suggests an error of $\sim$ 4.0% in the roll load is permissible.

To calculate the power requirements of the mill allowing for roll flattening:

$F_{R'}$ = 27.14 MN and R' = 444.9 mm

from equation (3.16) the corrected roll torque

$$= 0.475 \times 27.14 \times 10^6 \times \frac{172}{183} \sqrt{\frac{444.9 \times 16}{10^6}} \text{ N-m}$$
$$= 10.2 \times 10^5 \text{ N-m}$$

from equation (3.17a) the corrected power requirement

$$= 2 \times 2\pi \times 10.2 \times 10^5 \times \frac{66}{60} \text{ W}$$

$$= \underline{141 \times 10^5 \text{ W}}$$

Thus in this case when roll flattening is allowed for the power requirements are increased by 3.7%.

Note on the calculation of Load, Torque and Power Requirements when hot rolling steel

As seen from examples 3.2 - 3.5, the procedures necessary for these calculations can be rather lengthy. Cook and McCrum (1958) have published a graphical presentation of data for 15 different steels at temperatures from 900-1200°C. Provided therefore that the steel composition and required reduction conditions lie within their data, their publication enables a much faster calculation of load and torque requirements to be made.

3.6 Effect of changes in stock temperature in the roll gap

During deformation the temperature of the stock may alter due to heat lost by conduction to the relatively cold work rolls and due to heat gained from deformation. The heat lost to the atmosphere from the side faces of the stock, in the short time while in the roll pass, will be so small that the effect on the overall temperature can safely be ignored.

(i) The temperature rise due to deformation, ΔT_D

If we assume that the total power required for deformation, w_D, is converted into heat (ignoring the very small amount of energy which will remain stored in the deformed structure), then the temperature rise in the stock due to deformation, ΔT_D, is given by:

$$\Delta T_D = \frac{w_D \, t_c}{V \rho_s s_s} \tag{3.19}$$

where V is the volume of metal in the roll gap, ρ_s and s_s are respectively the density and specific heat of the stock at the rolling temperature and t_c is the contact time, i.e. the time for material to pass from the entry to the exit plane in the roll gap.

(ii) The temperature drop due to conduction to the rolls, ΔT_K

If we assume perfect contact between rolls and stock, i.e. an infinite heat transfer coefficient then:

$$\text{Rate of Heat Loss} = K_s A (T_s - T_c)\sqrt{\frac{\rho_s \, s_s}{\pi K_s t_c}} \tag{3.20}$$

where A = contact area, T_s = initial stock temperature, T_c = instantaneous contact temperature, which is assumed to remain constant (see below) and K_s is the thermal conductivity of the stock (Hollander, 1970).

Integrating equation 3.20 with respect to time gives the total heat loss ΔQ_K where:

$$\Delta Q_K = 2A(T_s - T_c)\sqrt{\frac{\rho_s s_s K_s t_c}{\pi}} \tag{3.21}$$

Hence the temperature drop due to conduction of heat to the rolls, ΔT_K, is given by

$$\Delta T_K = \frac{\Delta Q_K}{V \rho_s s_s} \qquad (3.22)$$

T_c can be found from the relationship that:

$$T_c = \frac{T_s \sqrt{K_s \rho_s s_s} + T_R \sqrt{K_R \rho_R s_R}}{\sqrt{K_s \rho_s s_s} + \sqrt{K_R \rho_R s_R}} \qquad (3.23)$$

where the subscripts R refer to the work roll material.

In the calculation of ΔT_K an infinite heat transfer coefficient was assumed in equation 3.20. On the basis of this assumption Hollander (1970) compared theoretical heat losses with experimental values and showed that the actual losses were only 0.6 of the calculated values. This implies that a more realistic value of the temperature drop is 0.6 ΔT_K. The overall temperature change of the stock, ΔT, is therefore given by:

$$\Delta T = \Delta T_D - 0.6\ \Delta T_K \qquad (3.24)$$

This value can now be used to give a better estimate of the true temperature of the stock during deformation, which in turn can be used to redetermine the flow stress of the metal and hence give an adjusted value of roll load.

For a complete analysis of the temperature changes in rolling a finite element method, which requires computing techniques, would be necessary but for most practical purposes this approximation is adequate.

In calculating the rolling load a mean value, $\bar{\sigma}'$, of the plain strain flow stress for the steel has been used. This value is a function of the mean strain, the mean strain rate and also the mean temperature in the roll gap. The mean temperature of the stock in the roll gap, $\bar{T}$, is given by:

$$\bar{T} = \bar{T}_o + \frac{\Delta T}{2} \qquad (3.25)$$

where $\bar{T}_o$ is the mean temperature of the stock at entry. $\bar{T}_o$ is difficult to determine with accuracy as allowance must be made for heat losses prior to entry into the roll gap. At the high temperatures, (1200-900°C), for hot rolling steel these losses are due mainly to radiation to the surroundings and conduction to the run-out table. Their calculation is complicated by factors such as the state and thickness of the oxide scale. Thus although a temperature gradient exists in the stock prior to entry into the rolls it is difficult to quantify with any accuracy and is therefore neglected in the example below.

In the case of multi-pass rolling situations the determination of $\bar{T}_o$ for each pass is obviously still more complex.

Example - Using the rolling data given in the example in section 3.2, determine the true temperature and hence the change in rolling load if, at the temperature of working, ρ_s = 7550 kg/m^3, s_s = 656 J/kg °C and K_s = 27.45 W/m°C.

(a) To calculate ΔT_D

The torque, Γ, required on each roll for the pass is (from section 3.4) 9.86×10^5 N-m.

$\therefore$ From equation (3.17) $w_D = 2 \times 2\pi \times 9.86 \times 10^5 \times \frac{66}{60}$ W.

$= 13.63 \times 10^6$ W

To calculate t_c:

The contact angle θ_o is given by equation (3.3)

$$\cos\theta_o = \frac{R - \frac{\Delta h}{2}}{R} = \frac{424.5}{432.5}$$

$$\therefore \theta_o = 11.04^\circ$$

Since the rolls rotate at 66 revolutions/min and conditions of sticking friction are assumed,

$$t_c = \frac{11.04}{360} \times \frac{66}{60} = \underline{0.0337 \text{ s}}$$

The volume of metal being deformed, V, must now be calculated.

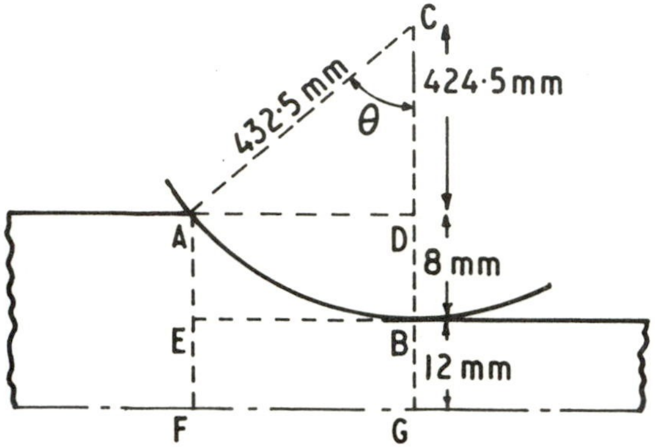

From the diagram:

$$V = \text{stock width} \times 2\ (\text{ABGF})$$

$$\text{ABGF} = \text{EBGF} + \text{ADBE} - (\text{ABC}-\text{ADC})$$

$$= 993.9 + 662.6 - (18021.4 - 17578.8)$$

$$= 1214 \text{ mm}^2$$

$$\therefore V = 1250 \times 2(1214) \text{ mm}^3$$

$$= \underline{3.035 \times 10^{-3} \text{ m}^3}$$

∴ From equation (3.19)

$$\Delta T_D = \frac{13.63 \times 10^6 \times 3.37 \times 10^{-2}}{3.035 \times 10^{-3} \times 7550 \times 656}\ ^\circ C$$

$$= \underline{30.6^\circ C}$$

(b) <u>To calculate ΔT_K</u>

For steel work rolls initially at 20°C,

$K_R = 48.13\ W/m^oC$, $s_R = 481.3\ J/kg^oC$ and

$\rho_R = 7840\ kg/m^3$.

By substitution in equation (3.23):

$T_c = 474.6^oC$

The length of the arc of contact $= \frac{\theta_o}{360} \times 2\pi R = \frac{11.04}{360} \times 2\pi \times 432.5$ mm

$= \underline{83.3 \text{ mm}}$

$\therefore$ The contact area $A = 2 \times \frac{83.3 \times 1250}{10^6}\ m^2$

$= \underline{0.208\ m^2}$

$\therefore$ From equation (3.21)

$$\Delta Q_K = 2 \times 0.208(1000-474.6)\sqrt{\frac{7550 \times 656 \times 27.45 \times 0.0337}{\pi}}\ J$$

$$= \underline{2.64 \times 10^5\ J}$$

$\therefore$ From equation (3.22)

$$\Delta T_K = \frac{2.64 \times 10^5}{3.035 \times 10^{-3} \times 7550 \times 656}\ ^oC$$

$$= \underline{17.6^oC}$$

Thus from equation (3.24)

$$\Delta T = 30.6 - 0.6(17.6)^oC$$

$$= \underline{20^oC}$$

Interpolation of the data of Cook and McCrum (1958) shows that the decrease of rolling load produced by this rise in temperature of the stock is very small, namely 2%. However, for lower rolling speeds and thinner stock the temperature changes can have a significant effect.

3.7 Spread during Hot Rolling

During a hot rolling pass a certain amount of spread, ie deformation parallel to the roll axes, may occur. This is more likely when rolling slab or plate than when rolling strip as the thickness to width ratio is larger in the former cases. The amount of spread occurring is usually quantified in terms of the spread coefficient S, where:

$$S = \frac{\ln \frac{w_f}{w_o}}{\ln \frac{h_o}{h_f}} \qquad (3.26)$$

Thus when S = 0 all the deformation is in the rolling direction and when S = 1 all the deformation would be in the width direction, ie for both situations the deformation is plane strain. When S = 0.5 the width and longitudinal strains are equal and thus the flow stress of the material will be that for axi-symmetric compression.

To calculate S, which increases with increase in $\frac{R}{h_o}$ and with decrease in Δh and $\frac{w_o}{h_o}$, a number of different relationships have been proposed (Sparling 1977). When $\frac{w_o}{h_o}$ lies between 3 and 16 the relationship proposed by Beese (1972) seems to be the most satisfactory, namely that:

$$S = 0.61\left(\frac{h_o}{w_o}\right)^{1.3} \exp\left\{-0.32\left(\frac{h_o}{\sqrt{R\Delta h}}\right)\right\} \qquad (3.27)$$

Provided therefore that the geometry of the deformation is such that $16 > \frac{w_o}{h_o} > 3$ use of equation 3.27 enables S to be calculated.

Both equations (3.26) and (3.27) assume that the initial cross section of the stock is rectangular and that the final stock width, w_f, is measured at the maximum width point, ie at the mid-thickness of the stock, if significant barrelling of the edges occurs. Additionally the measured spread should be far enough away from both the front and back end of the stock for "end spread" to be insignificant. Although verification of the above equations has been conducted on low carbon steel rolled at temperatures above 1000°C, it has been pointed out that neither the temperature, nor the strain rate, have a major effect on the amount of spread produced (Beese 1972).

Equation 3.27 involves the undeformed roll radius, and the equation has been determined using this value. It is not expected that roll flattening will make a significant contribution to the amount of spread occurring.

Example - Mild steel is rolled at 1000°C from 140 mm to a thickness of 89 mm on a plate mill with 910 mm diameter work rolls. If the initial width of the stock is 600 mm, determine the width remote from the ends after this single-pass reduction.

$$\frac{w_o}{h_o} = \frac{600}{140} = 4.29$$

Thus, with this width to thickness ratio, equation (3.27) is valid. Substitution in (3.27)

$$S = 0.61 \left(\frac{140}{600}\right)^{1.3} \exp\left\{-0.32 \frac{140}{\sqrt{455 \times 51}}\right\}$$

$$S = 0.61 \times 0.151 \times 0.745 = \underline{0.0686}$$

Now in equation (3.26)

Rearranging:

$$w_f = w_o \exp\left\{S \ \ln \frac{h_o}{h_f}\right\}$$

$$= 600 \exp\left\{0.0686 \ \ln\left(\frac{140}{89}\right)\right\}$$

Hence $w_f = 600 \times 1.032 = \underline{619 \text{ mm}}$

It is evident that the amount of spread is still only relatively small even though the width to thickness ratio is quite small.

3.8 Roll Load for Intermediate width stock $\left[2 < \frac{w_o}{h_o} < 6\right]$

For deformation conditions where spread is appreciable, the assumption made in the rolling theories, that plane strain conditions exist, is no longer valid. The actual separating force generated during rolling needs modification to account for the fact that the area of metal being rolled is greater than for plane strain, and the flow stress itself is changed. This modification has been carried out by Helmi and Alexander (1969) and their equation for rolling load without spread (which gives results that are in reasonable agreement with Sim's, equation (3.8)), has an additional term added, as shown below. The first two terms do not involve spread, but the last one does.

$$\frac{F}{\bar{w}} = \bar{\sigma}' \left\{\frac{\pi}{4} L + \frac{L^2}{2(h_o+h_f)} - \left(\frac{L^3\bar{h}}{3\bar{w}(h_o+h_f)^2}\right)\left(1-0.215\frac{(h_o+h_f)^3}{L}\right)\right\} \quad (3.28)$$

where $L = \sqrt{R'\Delta h}$, and $\bar{w} = \frac{w_o+2w_f}{3}$ is the "mean" width of the rolled stock and $\bar{h} = \frac{h_o+2h_f}{3}$ is the "mean" thickness.

In this equation the flow stress that should be substituted is the value for plane strain, the load correction for spread arising only from the last term inside the brackets. It should be noted in passing that the publication by Helmi and Alexander has a typographical error and the correct equation should be as laid out above (Alexander 1982).

Example - For the conditions of the previous example, determine the new rolling load taking account of spread and compare this with the load predicted for plane strain conditions. It may be observed for these conditions that the average plane strain flow stress is 180 N/mm^2.

Equation (3.28) involves the deformed roll radius. Thus a first estimate of the rolling load without spread and based on radius 455 mm is obtained from the first two terms of equation (3.28). This gives 30 647 N/mm width. Proceeding as in section 3.5, an improved value of roll radius can be obtained from Hitchcock's formula. Substituting in equation (3.18) for steel rolls gives the deformed roll radius as 462.1 mm.

With this value of roll radius, the true load without spread may be obtained again from the first two terms of (3.28)

$$\frac{F}{\bar{w}} = 180\left\{\frac{\pi}{4}\sqrt{462.1 \times 51} + \frac{462.1 \times 51}{2 \times (140+89)}\right\}$$

$$= 180 \times 172.0 = \underline{\underline{30\ 963 \text{ N/mm width}}}$$

The rolling load including spread can be found from (3.28) in a similar way.

The mean width of stock is, from section 3.8, given by:

$$\bar{w} = \frac{600 + 2 \times 619}{3} = 612.7 \text{ mm}$$

and $$\bar{h} = \frac{140 + 2 \times 89}{3} = 106 \text{ mm}$$

Hence $$\frac{F}{\bar{w}} = 180\left\{172.0 - \frac{(462.1\text{x}51)^{\frac{3}{2}}(106)}{3\text{x}612.7(229)^2}\left[1 - 0.215\frac{229^3}{(462.1\text{x}51)^{\frac{3}{2}}}\right]\right\}$$

$$\frac{F}{\bar{w}} = 180\{172.0 - 1.14\} = 180 \times 170.9$$

$$= \underline{\underline{30\ 755 \text{ N/mm width}}}$$

Hence the total rolling load is

$$\text{without spread} : \frac{30\ 963 \times 600}{9.81 \times 10^3} = \underline{\underline{1894}} \text{ tonne}$$

$$\text{with spread} : \frac{30\ 755 \times 612.7}{9.81 \times 10^3} = \underline{\underline{1921}} \text{ tonne}$$

It is evident from the result that the total load is large: when spread is taken into account. This is clearly a resu of the width of stock being greater, since the load/unit width is, in fact, smaller. Thus we have two compensatory effects; the increase in load due to increased area, and the decrease due to the modification of the friction hill on which the equation (3.28) is based (Helmi and Alexander 1969).

3.9 Roll Load allowing for Flat rolling of Billet Shaped Stock $\left(1 \leq \frac{w_o}{h_o} \leq 2\right)$

When the width of the ingoing stock is of approximately the same dimension as the material thickness then the flow stress of the material is less than the plane strain flow stress. It is however possible to make an estimate of the correct flow stress in terms of the coefficient of spread S.

$$\bar{\sigma}_{eff} = \bar{\sigma}' \sqrt{1 - S + S^2} \qquad (3.29)$$

where $\bar{\sigma}_{eff}$ is the effective flow stress of the metal. The form of the equation is such that when S = 0 the flow, along the direction of rolling gives the flow stress as σ' and also when S = 1 with flow parallel to the axis of the rolls the flow stress is σ'. When S = ½ the equation suggests that the flow stress is that for uniaxial compression as required.

Experimental investigation of rolling loads for this geometry of product, shows that the loads could be determined very satisfactorily for all the values of $\frac{R'}{h_f}$ and r examined. These ranged from 3.75 to 10 for the former and from 10 - 30% for the latter (Helmi & Alexander 1969).

As in the previous section the deformed roll radius is required. This is again affected by the effective flow stress of the metal. Equation (3.27) enables a value of S to be found and then substitution in (3.29) will give $\bar{\sigma}_{eff}$. Again using the first two terms of (3.28) the load can be obtained for use with the roll flattening equation (3.18). With an improved value of radius the procedure is exactly the same as in the previous example.

Example - If the width of the ingoing stock in the previous example is reduced to 210 mm, determine the rolling load expected.

For these dimensions $\frac{w_o}{h_o} = \frac{210}{140} = 1.5$

Hence the equation (3.29) must be used to determine the effective flow stress.

The amount of spread S is given from equation (3.27) independent of the flow stress.

$$S = 0.61\left(\frac{140}{210}\right)^{1.3} \quad \exp\left(\frac{-0.32 \times 140}{\sqrt{455 \times 51}}\right)$$

$$S = 0.268$$

and from equation (3.26):

$$w_f = 210 \exp\left(0.268 \ \ln \frac{140}{89}\right) = 237.1 \text{ mm}$$

The value of S will now give the effective stress from (3.29)

$$\bar{\sigma}_{eff} = 180\left\{1 - 0.268 + (0.268)^2\right\}^{\frac{1}{2}}$$

$$\bar{\sigma}_{eff} = 161.4 \text{ N/mm}^2$$

To obtain the improved value of roll radius the undeformed arc of contact L is needed.

$$\text{Thus } L = \sqrt{R\Delta h} = \sqrt{455 \times 51}$$

$$= 152.3 \text{ mm}$$

The first two terms in equation (3.28) now give

$$\frac{F}{\bar{w}} = 161.4\left\{\frac{\pi}{4} 152.3 + \frac{(152.3)^2}{2 \times 229}\right\}$$

or $\frac{F}{\bar{w}} = 27480$ N/mm width

Again, as previously,

$$R' = 455\left(1 + \frac{2.6 \times 10^{-5} \times 27480}{51}\right)$$

$$R' = 461.4 \text{ mm}$$

Although this radius of the deformed roll is calculated assuming no spread, and it is likely to be slightly modified when spread occurs (since load/unit width enters Hitchcock's Formula), it is the best estimate that can be made. (A further iteration for R makes less than 1% change in load/unit width).

The average width is now

$$\bar{w} = \frac{210 + 2 \times 237.1}{3} = 228.1 \text{ mm}$$

The average thickness $\bar{h} = \dfrac{140 + 2 \times 89}{3} = 106$ mm

and the deformed arc of contact

$$L = \sqrt{461{,}4 \times 51} = 153.4 \text{ mm}$$

Substitution in equation (3.28) gives

$$\frac{F}{w} = 161.4\left\{\frac{\pi}{4}(153.4) + \frac{(153.4)^2}{2x229} - \left\{\frac{(153.4)^3 \times 106}{3x228.1x(229)^2}\right\}\left\{1 - \frac{0.215x229}{153.4}\right\}^3\right\}$$

$$= 161.4\ \{120.5 + 51.38 - 7.24\}$$

$$\frac{F}{w} = 26573 \text{ N/mm}$$

Hence total load F = <u>617.9</u> tonne.

The reduction chosen in this example lies outside the range of reductions for which equation (3.28) has been verified experimentally. However, there is good agreement between the predicted and experimental loads for all reductions up to 30% when the ratio $\frac{R'}{h_f}$ is less than about 6; and so it would be expected that the load predicted above should be reasonably accurate. The agreement with experiment becomes progressively less satisfactory as the reduction increases and $\frac{R'}{h_f}$ increases.

SECTION 4 - FORGING

Industrial forging processes are too complex to analyse because they frequently involve deformation of metal between shaped dies. Furthermore, many of the operations are carried out without lubrication at elevated temperatures, so simple sliding friction conditions between the workpiece and tools no longer apply and the flow stress of the material is sensitive to temperature and strain rate, which may vary from place to place in the workpiece. Slip line field solutions are available for some complex geometries, but do not take account of the other variables. Solution of realistic problems therefore frequently requires sophisticated computing procedures. Both these methods are beyond the scope of this book.

The examples in this section have therefore been chosen for their geometrical simplicity to enable analytical solutions to be obtained. However, they illustrate important principles and serve as a foundation for consideration of more commercially interesting problems.

4.1 Determination of the Coefficient of Friction in Upset Forging

The concept of a coefficient of friction is used widely to express the surface resistance to motion of the workpiece relative to the tools. The greater the normal pressure at the interface the greater is the shear stress opposing motion. This can be expressed as:

$$\tau = \mu p \qquad (4.1)$$

The constant of proportionally μ varies widely with the type of lubricant, the surface conditions and environment, but also it can vary locally due to local changes in speed of deformation or temperature conditions. However, in practical working operations it is convenient to define an average coefficient that applies to the deformation of the whole workpiece within the die. An accurate estimate of its value is essential for prediction of loads and metal flow.

The coefficient of sliding friction can be deduced from the change in geometry of a ring during axisymmetric compression (Male and Cockcroft 1964). Although the ring under investigation must have a similar geometry to the test pieces used by Male and Cockcroft, the method has the advantage that it applies over a wide range of strain rates, temperatures and lubricants and is therefore independent of the flow stress of the material under test.

For a given starting geometry and reduction, the material flow depends on the value of the coefficient of friction. For very small values of μ all flow, at any radius within the annulus, is radially outwards. As μ increases to a critical value of 0.055 the internal diameter of the ring increases less and less until at the critical value of μ it does not change during compression. At higher coefficients the internal diameter of the ring decreases, whilst the external diameter continues to increase.

These changes in diameter have been related to the friction coefficient for a wide range of working conditions and a set of calibration curves has been produced.

The percentage decrease in internal diameter, $\Delta D\%$, of the forged ring for friction coefficients greater than 0.055 is given by:

$$\Delta D\% = k \ \ell n \ (\mu/0.055) \qquad (4.2)$$

where

$$\ell n \ k = (0.044 \times \% \text{ Red. in Ht.}) + 1.06 \qquad {*}(4.3)$$

These equations hold provided that the initial shape of the test piece is constant, with an O.D./I.D. ratio of 2 and an O.D./Height ratio of 3. The deformation range investigated lies between 20-60% engineering strain which is typical of

---

* [There is a misprint in the equation given in the original publication.]

forging reductions. For conditions where the frictional coefficient may not be constant during the test, for example at very high reductions, the calibration curve no longer applies and assessment of μ should be made directly from load calculations (Schroeder and Webster 1949, Avitzur 1968).

Example - An aluminium ring of initial dimensions 60 mm O.D., 30 mm I.D. and 20 mm in height, is upset forged at room temperature to a height of 12 mm, without lubrication. If the hole diameter closes during compression to 23.4 mm, determine the average coefficient of friction during the stroke.

$$\% \text{ Change in Diameter} = \frac{(30 - 23.4)}{30} \times 100 = 22\%$$

$$\% \text{ Reduction in Height} = \frac{(20 - 12)}{20} \times 100 = 40\%$$

From equation (4.3)

$$\ell n\ k = (0.044 \times 40) + 1.06$$

$$\ell n\ k = 1.76 + 1.06 = 2.82,$$

$$k = 16.78$$

Rearranging equation (4.2)

$$\mu = 0.055 \exp\left(\frac{22}{16.78}\right)$$

$$\mu = \underline{0.204 = 0.20}$$

Confidence in the result can only be to the second significant figure and then only if the material under test behaves in a similar manner to the material tested for the calibration curves over the strain range investigated. This method is therefore most useful in making comparisons rather than for the determination of fundamental values. The geometry of the test piece is so chosen that the zones of deformation from the upper and lower surfaces meet across the centreline, i.e. adequate penetration of deformation occurs.

Schey (1970) has emphasised the care that should be exercised in viewing the results. He points out that in the ring test there is very limited sliding between tool and workpiece, and this can lead to an optimistic view of lubricants that are useless for other purposes.

4.2 Stresses Required for Upset Forging of Solid Cylinders

The stresses developed in the upset forging of a cylindrical part in the presence of friction has been discussed in Chapter 1, equation (1.14), and of course, if the average flow stress were known, then the forging load could be determined. This approach uses the coefficient of friction to express the degree of tool-workpiece interaction. This is satisfactory where this is small and sliding occurs. However, when sticking at the interface occurs the equations are no longer valid.

Thus many engineering texts prefer to view interfacial friction as independent of the normal pressure and consider the surface resistance to motion as a constant factor, m, of the shear yield stress of the material.

This can be expressed as:

$$\tau = m\tau_0 = \frac{m\sigma_0}{\sqrt{3}} \qquad (4.4)$$

where m can vary between 0 and 1.

When $m = 0$ there is no interfacial friction

When $m = 1$ friction is so large that the surface layers of the workpiece are sheared at the shear yield stress of the material, i.e. sticking friction occurs. This condition puts an upper limit on the shear stress in equation (4.1). When the normal pressure exceeds a certain level τ cannot become greater than the shear yield stress and it thus remains constant at this latter value. In general, when sticking does not occur we can write, from equations (4.1) and (4.4)

$$\mu = \frac{m\sigma_0}{p\sqrt{3}} \qquad (4.5)$$

The concept of a friction factor m is useful in analytical derivations and in the calculation of upper bound solutions (Avitzur 1968). It is also applicable over the range of working conditions from sliding to sticking friction.

An upper bound solution to simple upset forging, using m, has been obtained on the assumption that barrelling does not occur. If a is the instantaneous cylinder radius and h is the height, then the average pressure is given by:

$$\frac{\bar{p}}{\sigma_0} = 1 + \frac{2m}{3\sqrt{3}} \frac{a}{h} \tag{4.6}$$

The form of this equation is identical with equation (1.14) but is valid over the whole range of frictional conditions. It is clear from equation (4.6) that the average pressure needed for deformation increases as a/h increases for constant friction conditions. This is a geometrical effect and indicates the increasing importance of the surface as the tools move together. It should be emphasised that it is not a work hardening effect since σ_0 in the equation can be considered constant.

Example - Two cylinders have the same radius, 100 mm, and heights of 100 mm and 50 mm. They are both upset forged to 50% of their original heights at elevated temperature by a ram travelling at a constant speed of 50 mm/s. The frictional factor for the compression is 0.9. The material obeys the relationship:

$$\sigma = 225|\varepsilon|^{0.3}|\dot{\varepsilon}|^{0.1} \text{ N/mm}^2$$

Determine both the average pressure and the load required for deformation as a function of ram travel.

The average pressure is given by equation (4.6) which involves the instantaneous height and radius. Since volume remains constant:

$$\pi a_o^2 . h_o = \pi a^2 h$$

and so

$$\frac{a}{h} = \frac{a_o}{h_o} \cdot \left(\frac{h_o}{h}\right)^{3/2}$$

and if x is the distance travelled by the ram then

$$h = (h_o - x)$$

Thus the relative pressure is

$$\frac{\bar{p}}{\sigma_0} = \left\{1 + \frac{2ma_o}{3\sqrt{3}\, h_0} \left(\frac{h_0}{h_0 - x}\right)^{3/2}\right\}$$

However, σ_0 is also a function of x through strain and strain rate

$$\text{strain } \varepsilon = \ln\left(\frac{h}{h_0}\right) = \ln\left(\frac{h_0 - x}{h_0}\right)$$

$$\text{strain rate } \dot{\varepsilon} = \frac{d\varepsilon}{dt} = \frac{dh}{h} \times \frac{1}{dt}$$

$$\text{or } \dot{\varepsilon} = \frac{v}{h} = \left(\frac{v}{h_0 - x}\right)$$

where v is the ram speed.

Substituting:

$$\bar{p} = 225 \left|\ln\left(\frac{h_0 - x}{h_0}\right)\right|^{0.3} \left|\left(\frac{v}{h_0 - x}\right)\right|^{0.1} \left\{1 + \frac{2ma_o}{3\sqrt{3}\, h_0} \left(\frac{h_0}{h_0 - x}\right)^{3/2}\right\}$$

for any value of x.

The mod symbols appear in the equation because, strictly, the strain is negative.

In a similar way the instantaneous area can be expressed:

$$A = \pi a_o^{\,2}\left(\frac{h_0}{h_0 - x}\right)$$

and so the instantaneous load is

$$L = \bar{p}A.$$

The pressure and the associated load have been calculated for different values of x and the results are tabulated and plotted in the graph, figure 4.1. The calculations are most conveniently done using a small computer.

	Initial Height 100 mm				
Ram Travel, x(mm)	10	20	30	40	50
New Height (mm)	90	80	70	60	50
Strain	0.1054	0.2231	0.3567	0.5108	0.6932
Average Pressure, (N/mm^2)	151.8	203.2	254.1	315.2	399.1
Load, (MN)	5.300	7.978	11.405	16.506	25.074

	Initial Height 50 mm				
Ram Travel, x(mm)	5	10	15	20	25
New Height (mm)	45	40	35	30	25
Strain	0.1054	0.2231	0.3567	0.5108	0.6932
Average Pressure, (N/mm^2)	209.7	288.8	373.6	482.1	639.4
Load, (MN)	7.320	11.339	16.77	25.245	40.17

For each forging the final areas and strains are the same, and yet the pressure required for the forging with the smaller height/diameter ratio is much greater than for the other one. This is due to the combined effect of friction and increased strain rate in the smaller forging.

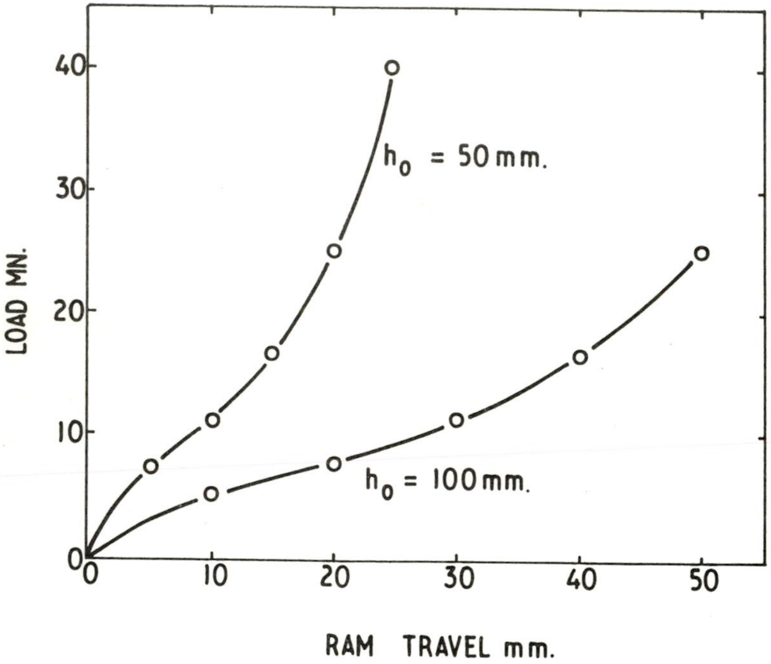
40
30
20
10
0
LOAD MN.
h_0 = 50 mm.
h_0 = 100 mm.
0
10
20
30
40
50
RAM TRAVEL mm.

Fig. 4.1

4.3 Press Forge Cogging

During cogging of bar, if the forging tool width, w, is greater than the thickness of the bar, h, the deformation approximates to plane strain compression. For sticking friction, in plane strain compression, it is simply shown (e.g. Rowe, 1965) that the force, F, required for deformation is given by:

$$F = wb.\ f\sigma_e \left(1 + \frac{w}{4h}\right) \tag{4.7}$$

where b is the breadth of the bar, σ_e is the equivalent tensile flow stress of the material and f is a factor to allow for the plane strain conditions. For true plane strain, i.e. b = constant so that $\varepsilon_2 = 0$, the factor f, from von Mises criterion, is $2/\sqrt{3} = 1.155$.

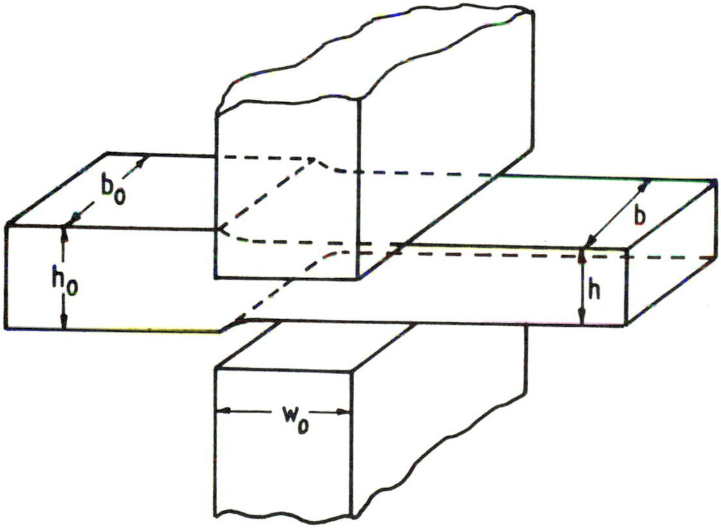

The strain imparted in the thickness direction, $\varepsilon_3 = \ell n\ h/ho$, is related to the equivalent tensile strain as (Nadai, 1949)

$$\varepsilon_e = \frac{\sqrt{2}}{3}\left[(\varepsilon_1 - \varepsilon_2)^2 + (\varepsilon_2 - \varepsilon_3)^2 + (\varepsilon_3 - \varepsilon_1)^2\right]^{\frac{1}{2}} \tag{4.8}$$

and because volume is constant so that $\varepsilon_1 + \varepsilon_2 + \varepsilon_3 = 0$,

$$\varepsilon_e = \frac{2}{\sqrt{3}} \left| \varepsilon_3 \right| = f \left| \varepsilon_3 \right| \qquad (4.9)$$

for plane strain conditions.

In practice some spread tends to occur, so $\varepsilon_2 = \ell n\, b/bo$ is real and must be measured.
Then,

$$\varepsilon_e = \frac{2}{\sqrt{3}} (\varepsilon_2^2 + \varepsilon_2\varepsilon_3 + \varepsilon_3^2)^{\frac{1}{2}} = f\, |\varepsilon_3| \qquad (4.10)$$

and the value of f can be found to substitute into equation (4.7).

Example - Calculate the force required to forge a bar of stainless steel 200 mm in thickness and 800 mm in breadth by 10% reduction in thickness if the forging tool is of width 250 mm (and a 'bite' of full tool width is taken) and the speed is constant at 50 mm/s, when (a) the breadth remains constant and (b) the breadth increases to 812 mm.

Assume temperature is constant at 1150°C and that at this temperature the equivalent tensile flow stress, σ_e, of the material is related to the equivalent tensile strain, ε_e, and strain rate, $\dot{\varepsilon}_e$, as

$$\sigma_e = 225\, \dot{\varepsilon}_e^{\,0.1}\, \varepsilon_e^{\,0.3} \quad N/mm^2$$

(a) In order to determine σ_e, it is first necessary to determine the equivalent strain and strain rate for plane strain deformation.

$$\varepsilon_3 = \ell n \frac{180}{200} = -0.105$$

Hence $$\varepsilon_e = \frac{2}{\sqrt{3}} \times 0.105 = 0.122$$

The strain rate of interest is the value just before the ram stops moving (which is taken to be instantaneous in this example), i.e.

$$\dot{\varepsilon}_3 = \frac{v}{h} = \frac{-50}{180} = -0.28\ s^{-1}$$

Hence $\quad \dot{\varepsilon}_e = \frac{2}{\sqrt{3}} \times 0.28 = 0.32\ s^{-1}$

and $\quad \sigma_e = 225(0.32)^{0.1}\ (0.122)^{0.3}$

$$= \underline{107\ N/mm^2}$$

From equation (4.7)

$$F = 250 \times 800 \times \frac{2}{\sqrt{3}} \times 107 \left(1 + \frac{250}{4 \times 180}\right)$$

$$= \underline{33.3\ MN}$$

(b) $\quad \varepsilon_3 = -\ 0.105$

$$\varepsilon_2 = \ell n\ \frac{812}{800} = 0.0149$$

Thus, from equation (4.10)

$$\varepsilon_e = 0.114$$

and $\quad f = \frac{0.114}{0.105} = 1.082$

Hence $\quad \dot{\varepsilon}_e = f|\dot{\varepsilon}_3| = 1.082 \times 0.28$

$$= 0.30\ s^{-1}$$

and $\quad \sigma_e = 225(0.30)^{0.1}\ (0.114)^{0.3}$

$$= \underline{104\ N/mm^2}$$

From equation (4.7)

$$F = 250 \times 812 \times 1.082 \times 104 \left(1 + \frac{250}{4 \times 180}\right)$$
$$= \underline{30.8\ MN}$$

Thus, despite the increase in breadth, the force is reduced by 7½% because of the relaxation of the plane strain constraint.

4.4 Upset Forging Under a Drop Hammer

In drop hammering, a tup of mass, M, falls freely to attain a velocity, v_o, on impact with the workpiece. It is then brought to rest by deforming the workpiece. The instantaneous force, F, exerted depends on the instantaneous rate of change of velocity and is thus given by:

$$M \frac{dv}{dt} = -F \qquad (4.11)$$

The value of F depends on the flow stress of the material, σ, as a function of the strain, ε, strain rate, $\dot{\varepsilon}$, and temperature, T, on the cross sectional area, A, and on the frictional conditions.

For upsetting of a cylinder, if homogeneous, frictionless deformation is assumed, then

$$F = \sigma \frac{A_o h_o}{h} \qquad (4.12)$$

where h_o and h are the original and instantaneous heights. With this assumption, and flow stress related only to strain rate as

$$\sigma = K\dot{\varepsilon}^m \qquad (4.13)$$

Johnson and Mellor (1973) obtained an algebraic solution to the differential equation (4.11) given by

$$M\left(\frac{v_o^{2-m} - v^{2-m}}{2 - m}\right) = \frac{A_o h_o K}{m} \left(\frac{1}{h^m} - \frac{1}{h_o^m}\right) \qquad (4.14)$$

provided $m \neq 0$. By putting $v = 0$, this equation enables the height reduction per blow to be calculated. However, the assumptions are unrealistic for normal hot forging, when sticking friction conditions are likely to exist. The relationship for these conditions is given by (Rowe 1965)

$$F = \sigma \frac{A_o h_o}{h} \left(1 + \frac{2a}{3\sqrt{3}\,h}\right) \qquad (4.15)$$

if homogeneous deformation is still assumed and a is the instantaneous radius.

Furthermore over the usual strain range of interest the effect of work hardening should not be overlooked and equation (4.13) should be replaced, for example,by :

$$\sigma = K' |\dot{\varepsilon}|^m |\varepsilon|^n \tag{4.16}$$

Substitution of equations (4.15) and (4.16) into equation (4.11) leads to the relationship

$$M\left(\frac{v_o^{2-m} - v^{2-m}}{2 - m}\right) = A_o h_o K' \int_{h_o}^{h} \left(\ln \frac{h_o}{h}\right)^n \left[\frac{1}{h^{1+m}} + \frac{0.217(A_o h_o)^{0.5}}{h^{2.5+m}}\right] dh \tag{4.17}$$

where the right hand side must be integrated graphically or numerically.

<u>Example</u> - A drop hammer with a tup of 8 tonnes mass is to be used to upset forge a cylinder of type 316 stainless steel at 1150°C. If the cylinder is initially 200 mm diameter and 200 mm high and the tup is dropped through a distance of 2 m calculate:

(i) the reduction in height achieved in one blow:

(a) by applying equation (4.14) with

$K = 105$ MN/m² and $m = 0.1$

(b) by applying equation (4.17) with

$K' = 225$ MN/m², $m = 0.1$ and $n = 0.3$

(ii) the number of blows required to achieve a 50% reduction in height if assumption (a) applies and temperature is assumed to remain constant

For a drop of x m,

$$v_o = (2\ gx)^{\frac{1}{2}}$$

$$= (2 \times 9.81 \times 2)^{\frac{1}{2}}$$

$$= \underline{6.26\ m/s}$$

(i) (a) From equation (4.14), when v = 0

$$8000 \times \frac{6.25^{1.9}}{1.9} = \frac{\pi(0.20)^2 \times 0.20 \times 105 \times 10^6}{4 \times 0.1} \left[\frac{1}{h^{0.1}} - \frac{1}{0.2^{0.1}}\right]$$

$$1.369 \times 10^5 = 6.597 \times 10^6 \left[\frac{1}{h^{0.1}} - 1.175\right]$$

$$\frac{1}{h^{0.1}} = 2.08 \times 10^{-2} + 1.175$$

$$h = \underline{0.168\ m}$$

$$\text{Reduction in height} = \frac{0.200 - 0.168}{0.200} \times 100$$

$$= \underline{16.1\%}$$

(i) (b) Transfer the constants in equation (4.17) to the left hand side and put v = 0, to give

$$\frac{M\, v_o^{2-m}}{2 - m} \cdot \frac{1}{A_o h_o K'} = \frac{5000 \times 6.26^{1.9}}{1.9} \times \frac{4}{\pi(0.2)^2 \times 0.2 \times 225 \times 10^6}$$

$$= \underline{6.07 \times 10^{-2}}\ (m^{-0.1})$$

Calculate the term inside the integral in equation (4.17) for a series of values of h decreasing in small increments and, for graphical solution, plot figure 4.2. From the area under the curve, plot figure 4.3.

Say, $J = \ln\left(\frac{h_o}{h}\right)^{0.3} \left[\frac{1}{h^{1.1}} + \frac{0.217\left(\frac{\pi}{4}(0.2)^2 \times 0.2\right)^{0.5}}{h^{2.6}}\right]$

By substitution

h, m	J $m^{-1.1}$	$\int_{h_o}^{h}$ Jdh (from figure 4.2)
0.200	0	0
0.199	1.439	-
0.195	2.405	8.94×10^{-3}
0.190	3.079	2.274×10^{-2}
0.185	3.620	3.954×10^{-2}
0.180	4.114	5.884×10^{-2}
0.175	4.592	8.054×10^{-2}

From figure 4.3, the value of h when $\int$Jdh is

$= 6.05 \times 10^{-2}(m^{-0.1})$ is

$h = \underline{0.1795 \text{ m}}$

Reduction in height = $\underline{10.3\%}$

Note that the two equations for σ give an equal value when $\varepsilon = 0.079$, i.e. part way through the blow, so the difference between the answers in (i)(a) and (b) arises mainly from the effects of friction, i.e. from the last terms in equations (4.15) and (4.17). Clearly, frictional effects are of great importance in forging.

(ii) After the first blow h = 0.168 m and for the second blow, this becomes h_o in equation (4.14). However, $A_o h_o$ is the volume of the workpiece and so remains constant for each blow. Proceeding as in part i(a), after the second blow:

$$1.369 \times 10^5 = 6.597 \times 10^6 \left(\frac{1}{h^{0.1}} - \frac{1}{0.168^{0.1}}\right)$$

$$h = 0.141 \text{ m}$$

after the third blow

$$h = 0.119 \text{ m}$$

after the fourth blow

$$\underline{h = 0.101 \text{ m}}$$

Thus, four blows are required to give a 50% reduction in height (in the absence of frictional effects).

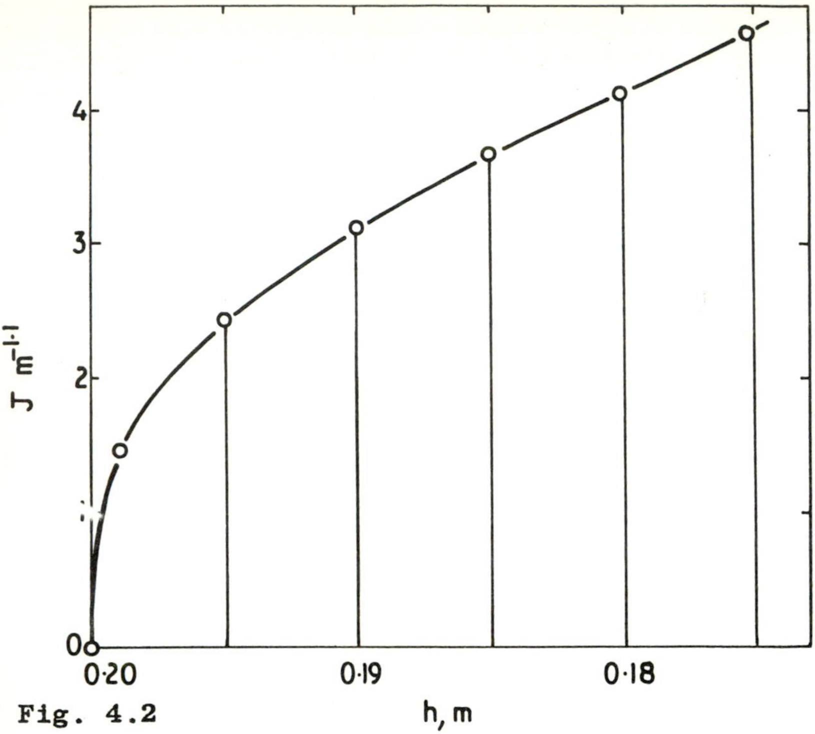

Fig. 4.2

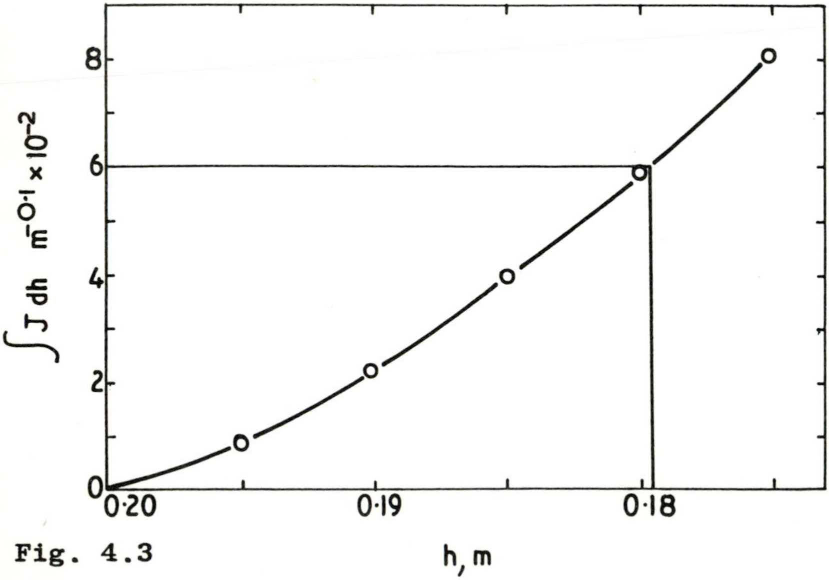

Fig. 4.3

4.5 Forging of Metal Powder Preforms

The prediction of forging loads associated with the deformation of powder preforms is complicated by the fact that the density of the material varies according to its past history, and therefore has a large influence on behaviour.

The final density achieved in simple compression of a preform depends on both the initial density and on the apparent strain given to the material. Since the total mass remains constant during the stroke, simple geometry shows that for compression of a cylinder without barrelling:

$$\rho_f = \left(\frac{a_o}{a_f}\right)^2 \frac{h_o}{h_f} \cdot \rho_o \tag{4.18}$$

where ρ_o, a_o and h_o, and ρ_f, a_f and h_f, refer to the initial relative density, radius and height, and final relative density, radius and height, respectively. It is often convenient to use relative density rather than the absolute value. This is defined as the ratio of the instantaneous density to the theoretical density of the fully dense material.

The ratio between lateral strain and longitudinal strain remains constant during elastic deformation of solid materials and is called Poisson's ratio. In a similar way this ratio has a different, but constant, value during plastic deformation, expressing the fact that the volume of the material remains essentially constant. During plastic deformation of powders, however, the ratio is no longer constant, and is a function of the *instantaneous* density, ρ, although it is independent of the initial density. A satisfactory relationship has been determined experimentally by Kuhn and Downey (1973), and is given by:

$$\nu = 0.5\,\rho^2 \tag{4.19}$$

ν is commonly called the Poisson's Ratio, and has the value 0.5 when densification is complete. This corresponds to the ratio used in conventional plasticity theory.

As a result of density changes the usual rules of deformation in the presence of a three dimensional stress system require modification. Kuhn and Downey (1971) proposed, and experimentally verified, a working yield criterion which is an extension of the one proposed by von Mises. They suggested that for a powder compact the function, f, remains constant at yield where:

$$f = \left[3J(2)' - (1 - 2\nu)J(2)\right]^{\frac{1}{2}} = \text{constant} \qquad (4.20)$$

$J(2) = -(\sigma_1\sigma_2 + \sigma_2\sigma_3 + \sigma_3\sigma_1)$ is the second invariant of the stress,

and $J(2)' = \frac{1}{6}\left\{(\sigma_1 - \sigma_2)^2 + (\sigma_2 - \sigma_3)^2 + (\sigma_3 - \sigma_1)^2\right\}$ is the

second invariant of the stress deviator.

Subscripts 1, 2 and 3 refer to the three principal stress directions. When the material is fully dense the above equation reduces to that proposed by von Mises, as expected.

(a) Simple Upsetting with Friction

The criterion can be applied to determine the stress distribution across the tool-workpiece interface in simple axisymmetric forging and hence the load can be found. The normal pressure p, at any radius a, from the axis of a forged cylindrical preform can be shown to be:

$$\frac{p}{\sigma_0} = \frac{2\nu q}{\sigma_0} + \left\{1-2\left(\frac{q}{\sigma_0}\right)^2(1 - \nu - 2\nu^2)\right\}^{\frac{1}{2}} \qquad (4.21)$$

where q, the internal radial stress, created by friction at the tool interface, is given by

$$q = 2m\tau_0\left(\frac{a_o - a}{h}\right) \qquad (4.22)$$

$m\tau_0$ is defined in section 4.2, equation (4.4) and a_o and h are the initial outer radius and height of the cylinder,respectively.

The expression for relative normal pressure, obtained by combining equations (4.21) and (4.22), shows an increase in pressure from the surface of the cylinder to a maximum at the centreline i.e. it gives a friction hill similar to those produced in rolling (section 2.1). Summing the pressure over the area of compression gives the load necessary to give deformation.

$$\text{Total Load } L = \int_{a=o}^{a=a_o} 2\pi a p(a) da = 2\pi\sigma_0 \int_{a=o}^{a=a_o} \frac{p(a)}{\sigma_0} a.da \qquad (4.23)$$

Example (a) - A cylindrical compact of sintered sponge iron powder, height 50 mm, diameter 75 mm and initial relative density 0.84, is compressed under carefully controlled (frictionless) conditions to a height of 33 mm and diameter 88.7 mm. The final load is 207.9 tonne. Determine the relative normal pressure at any radius from the centre of the cylinder, and hence evaluate the new forging load, if the workpiece were subsequently forged in the presence of interfacial friction where the frictional factor *m* is 0.3.

The uniaxial yield stress of the compact after forging homogeneously to a height of 33 mm is given immediately:

$$\sigma_0 = \frac{\text{Load}}{\text{Area}} = \frac{207.9 \times 10^3 \times 9.81 \times 4}{\pi \times (88.7)^2} \quad \text{N/mm}^2$$

$$\sigma_0 = \underline{330 \text{ N/mm}^2}$$

In order to find the interfacial pressure, a value of ν corresponding to the final density is needed.

From equation (4.18)

$$\rho_f = \left(\frac{75/2}{88.7/2}\right)^2 \frac{50}{33} \times 0.84$$

$$\rho_f = \underline{0.91}$$

and using equation (4.19)

$$\nu = 0.5\ (0.91)^2$$

$$\underline{\nu = 0.414}$$

and the term

$$(1 - \nu - 2\nu^2) = 0.2432$$

Also combining equations (4.21) and (4.22) and remembering that according to von Mises $\tau_0 = \sigma_0/\sqrt{3}$ we can write

$$\frac{p}{\sigma_0} = \frac{4m(0.414)}{\sqrt{3}}\left(\frac{a_o - a}{h}\right) + \left[1 - \frac{8m^2(0.2432)}{3}\left(\frac{a_o - a}{h}\right)^2\right]^{\frac{1}{2}}$$

or for $m = 0.3$

$$\frac{p}{\sigma_0} = 0.2868\left(\frac{a_o - a}{h}\right) + \left[1 - 0.05837\left(\frac{a_o - a}{h}\right)^2\right]^{\frac{1}{2}}$$

Values of relative pressure have been calculated for different values of radius and are tabulated below and plotted in the diagram, figure 4.4

$a_o = 44.35$ mm, $h = 33$ mm

a	0	$0.1\ a_o$	$0.2\ a_o$	$0.3\ a_o$	$0.4\ a_o$
$\frac{a_o - a}{h}$	1.3439	1.210	1.0752	0.9408	0.8064
$\frac{p}{\sigma_0}$	1.331	1.303	1.274	1.244	1.212
$\frac{ap}{\sigma_0 a_o}$	0	.130	.244	.372	.484

a	$0.5\ a_o$	$0.6\ a_o$	$0.7\ a_o$	$0.8\ a_o$	$0.9\ a_o$	a_o
$\frac{a_o - a}{h}$	0.6720	0.5376	0.4032	0.2688	0.1344	0
$\frac{p}{\sigma_0}$	1.180	1.146	1.111	1.075	1.038	1
$\frac{ap}{\sigma_0 a_o}$	.590	.690	.777	.860	.936	1

The total load is obtained from equation (4.23). It is perhaps most convenient to evaluate the integral from a graph of $\left(\frac{ap}{\sigma_0}\right)\cdot\frac{1}{a_o}$ versus $\frac{a}{a_o}$ as shown in figure 4.5.

The area A, obtained graphically, can then be used to give the load from

$$L = 2\pi a_o^{\,2}\sigma_0 \cdot A$$

From the graph A = 0.559

$$\therefore \quad L = 2\pi\left(\frac{88.7}{2}\right)^2 \times \frac{330 \times 0.559}{9.81 \times 10^3}$$

or $\quad$ Load = 232.4 tonne

(b) Repressing without Friction

Repressing occurs when a compact has been forged by uniaxial compression past the point where the sides of the die exert a restraining influence through inhibition of further lateral flow. When this happens the stress required for further deformation increases due to the added restraint. In the same way as for a solid material, where von Mises criterion can be used to predict the stress, so the modified relationship (4.20) can be applied to find the new forging stress and load during repressing. In this case the complicating effects of friction at the tools and die walls have been ignored.

For simple cylindrical geometry where the radius and circumferential strains are maintained at zero, the criterion gives:

$$f = \sigma_0 = \sigma_1\left[\frac{(1 - 2\nu)(1 + \nu)}{(1 - \nu)}\right]^{\frac{1}{2}} \qquad (4.24)$$

It must be remembered that σ_0, the flow stress of a homogeneously deformed compact, is a function of the instantaneous density of the compact, and that this in turn depends on the prior history of the part. It is essential, therefore, that in order to determine the stress required for repressing of a compact of given density, the value of σ_0 used corresponds to that associated with the correct density history. The reason for this is that the degree of work hardening of the metal within the compact, as well as the degree of porosity, controls the strength. However, in applying equation (4.24) above there is an assumption that the work done on the metal in achieving a certain density is the same whether by simple compression or by any other mode of deformation.

Although these equations for the deformation of powders have been developed and tested on compacts at room temperature, Kuhn and Downey (1971) have pointed out that the theory should also be applicable for any material and at other temperatures.

Example (b) - If after compression to a height of 33 mm and diameter of 88.5 mm, the compact in the previous example came into contact with the die wall and the part were then repressed to a final height of 32 mm, determine the axial load required to achieve this size, given that the uniaxial flow stress corresponding to the final density is 380 N/mm^2.

Reduction in height from 33 mm to 32 mm at constant diameter gives a final relative density of

$$\rho_f = 0.91 \times \frac{33}{32} = 0.938$$

This density of compact has a uniaxial flow stress of 380 N/mm^2, provided that it has been deformed from an initial relative density of 0.84 (see previous example).

The associated Poisson's Ratio for this density is therefore, equation (4.19)

$$\nu = 0.5\ (0.938)^2 = 0.440$$

Thus substituting σ_0 and ν into equation (4.24)

$$380 = \sigma_1 \left[\frac{(1 - 2 \times 0.44)(1 + 0.44)}{(1 - 0.44)} \right]^{\frac{1}{2}}$$

or $\quad \sigma_1 = 684 \text{ N/mm}^2$

The repressing load in the absence of friction is therefore

$$L = \frac{\pi (88.5)^2}{4} \times \frac{684}{9.81 \times 10^3} \text{ tonne}$$

$$L = \underline{429 \text{ tonne}}$$

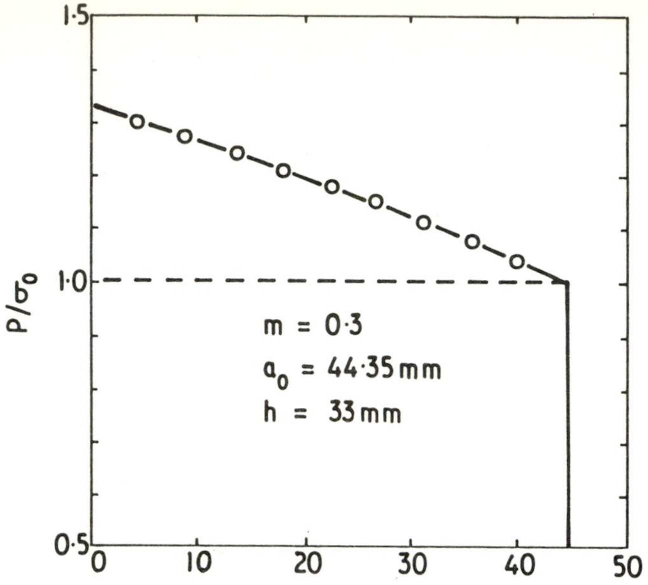

Fig. 4.4 DISTANCE FROM CENTRE (a) mm.

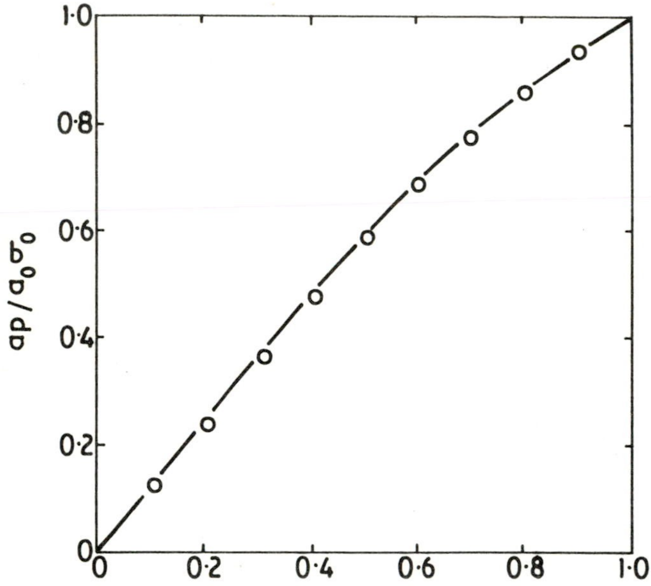

Fig. 4.5 a/a_0

4.6 Elastic Stresses Generated During Treatment of Large Forgings

When large forgings are heat treated (≥ 3 tonne for steel) then there is a temperature differential set up during the heating and cooling stage because the inside temperature lags behind the surface temperature. This in turn leads to the generation of internal stresses. If the temperature or stress is insufficient to allow any plastic deformation to occur then the purely elastic stresses will vanish once the temperature has equalised. If plastic flow occurs due to the stresses, then stress relaxation will occur at high temperature and residual stresses will be set up after temperature equalisation at low temperatures.

The temperature difference ΔT between the inside and the surface of a cylindrical forging during heating or cooling is given under steady state conditions by:

$$\Delta T = \frac{a_o^2}{4K} \frac{dT}{dt} \qquad (4.25)$$

where $\frac{dT}{dt}$ is the rate of heating (or cooling) and is the same for both the surface and the middle during steady state, a_o is the radius, in m, and $K = \frac{k}{\rho s}$ is the thermal diffusivity, in m^2/hr,

k is the thermal conductivity,

ρ is the density and

s is the specific heat.

The difference in temperature generates longitudinal, radial and circumferential stresses (Sykes 1953-54); the longitudinal being the greatest and is given by

$$\sigma_{max} = \frac{E\alpha}{2(1 - \nu)} \Delta T \qquad (4.26)$$

E is Young's Modulus, ν is Poisson's ratio and α is the coefficient of linear expansion of the material.

Combining the two equations above gives:

$$\sigma_{max} = \frac{E\alpha}{8K(1 - \nu)} a_o^2 \frac{dT}{dt} \qquad (4.27)$$

Thus if the physical constants for the material are known at different temperatures, it is possible to calculate the expected internal stresses for a particular forging for any given heating or cooling rate. This information is useful for the determination of permissible heating and cooling rates in large forgings so that fracture, distortion or residual stress do not ensue. The secondary advantage lies in the reduction in furnace heating and cooling times, and in allowing a forging to be brought into the air to cool without problems, instead of occupying a furnace at low temperatures unnecessarily.

In using the equations it must be remembered that the physical constants such as K, α, and E are temperature and composition dependent and hence the stresses calculated for a given temperature must take this into account. Thus the product $\frac{E\alpha}{K}$ increases by a factor of three during heating of a medium carbon, low alloy, steel from 0-700°C. This means that heating or cooling rates at the higher temperatures must be less than at lower temperatures if the level of internal elastic stresses are to be held constant and no stress relaxation occurs.

Below a certain temperature the internal stresses generated in air cooling may be less than the maximum permissible level of internal stress, and this allows removal from the furnace at an early stage.

The cooling rate in air per unit length of cylinder is

$$\frac{dT}{dt} = \frac{2}{\rho s a_o} F(T) \qquad (4.28)$$

where F(T) is the rate of heat loss by both convection and radiation per unit area of the cylinder. It is therefore possible to find the air cooling rate for different cylinder diameters provided that F(T) is known. Sykes (1953-54) has done this for oxidised steels using heat loss calculations made by Fishenden and Saunders (1934). These can be obtained from the data below for radius a_o.

Surface Temperature of Forging ^{o}C	100	200	300	400	500	600
$a_o \frac{dT}{dt}$ (mm ^{o}C h^{-1})	1905	5588	11938	19940	28320	38355

These data relating to air cooling refer to surface temperatures and these are significantly below the average temperatures for which the thermal data apply. It is therefore necessary to take this into account.

If a parabolic temperature distribution within the forging is assumed then it can be shown that during steady state cooling (Hughes and Sellars 1972)

$$T_s = \bar{T} - \frac{\Delta T}{2} \qquad (4.29)$$

Thus for a known cooling rate, equation (4.25) will enable ΔT, and hence the surface temperature, T_s, to be found for a given average temperature, $\bar{T}$. The cooling rates corresponding to these surface temperatures can now be obtained by interpolation from the tabulated data.

Example - A long cylindrical forging is 1800 mm diameter and made of medium carbon low alloy steel. It is fully annealed after forging and then cooled down to room temperature for machining. During cooling the instantaneous elastic stress levels must not exceed 100 N/mm^2 in the temperature range 600-350°C, and 150 N/mm^2 between 350°C and room temperature.

Determine:

(a) the maximum permissible cooling rates between 600°C and room temperature

(b) the temperature at which it is safe to remove the forging from the furnace

(c) the time of occupation in the furnace during cooling from 600°C.

The relevant data on physical constants are shown below.

Average Temperature ^{o}C	100	200	300	400	500	600
K m^2/hr	0.0394	0.03901	0.03784	.03589	.03316	.03004
$\alpha \times 10^{-6} (^{o}C)^{-1}$	12.4	13.7	14.7	15.5	16.3	16.9
E x 10^9 N/m^2	206.9	200.7	193.0	185.3	176.0	160.6
$E\alpha/K \times 10^6$	65.12	70.49	74.98	80.02	86.52	92.94

(a) Using these data, substitution into equation (4.27) and using a value of Poisson's ratio of 1/3 gives the cooling rate that will generate the maximum permissible stress.
For example, at 600°C:

$$\frac{dT}{dt} = \frac{8\left(1 - \frac{1}{3}\right)\sigma_{max}}{\frac{E\alpha}{K} \times a_o^{\ 2}}$$

$$= \frac{16}{3} \times \frac{100 \times 10^6}{92.94 \times 10^6} \times (0.9)^2$$

$$\frac{dT}{dt} = 7.08^{o}C/hr$$

i.e. a cooling rate faster than this will generate a stress greater than 100 N/mm^2

The maximum permitted cooling rates for all the temperatures are tabulated. Below 350°C the stress substituted in equation (4.27) is 150 N/mm^2.

Temperature ^{o}C	100	200	300	400	500	600
Permitted Stress MN/m^2	150	150	150	100	100	100
Permitted Cooling Rate $^{o}C/hr$	15.17	14.01	13.17	8.23	7.61	7.08

(b) From the table for air cooling rates the rate for the different surface temperatures can be found. These are plotted in figure 4.6 The actual values of T_s corresponding to the various $\bar{T}$ values must now be found.

If $\bar{T} = 600^{o}C$

a furnace cooling rate of $7.08^{o}C/hr$ must be imposed.

Thus in equation (4.25)

$$\Delta T = \frac{0.9^2 \times 7.08}{4 \times 0.03004}$$

$$= 47.7^{o}C$$

Then $T_s = 600 - 24 = 576^{o}C$

Similarly other values can be calculated for different $\bar{T}$.

$\bar{T}$ (^{o}C)	100	200	300	400	500	600
$\left(\frac{d\bar{T}}{dt}\right)_{Furnace}$	15.17	14.01	13.17	8.23	7.61	7.08
T_s (^{o}C)	61	163.6	264.7	376.8	476.8	576.1
$\left(\frac{d\bar{T}}{dt}\right)_{Air}$	$\sim$2	4.5	10.8	20.2	29.2	39.5
σ(N/mm^2)	19.8	48.2	123	245.5	383.7	557.6

From the surface temperature, the corresponding cooling rate *in air* can be read from figure 4.6. This in turn gives a stress generated by this rate, equation (4.27). The stresses developed by air cooling for a range of average forging temperatures are plotted in figure (4.7). Also indicated are the permitted stress levels in the different temperature ranges.

Inspection of the graph shows that below an average temperature of $325^{o}C$ the stresses generated during air cooling are less than the elastic stresses permitted.

Thus it is safe to remove the forging from the furnace once the average temperature has reached 325°C.

(c) Time in the furnace

$$t = \sum \frac{\text{Temperature Interval}}{\text{Cooling Rate}}$$

$$= \frac{100}{7.08} + \frac{100}{7.61} + \frac{75}{8.23}$$

$$t = \underline{36.4 \text{ hrs}}$$

Throughout this example it has been assumed that no stress relaxation occurs during cooling. The more complex problem where partial relaxation takes place has been discussed in detail by Sykes (1953-54.)

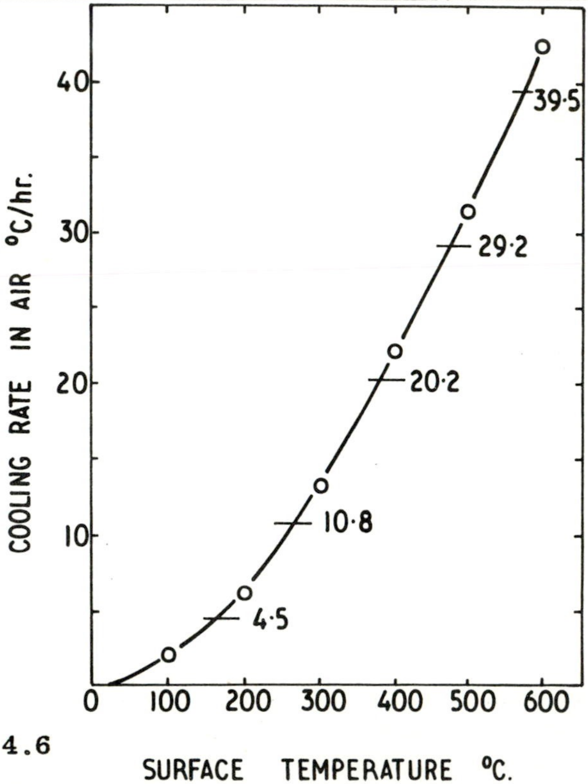

Fig. 4.6

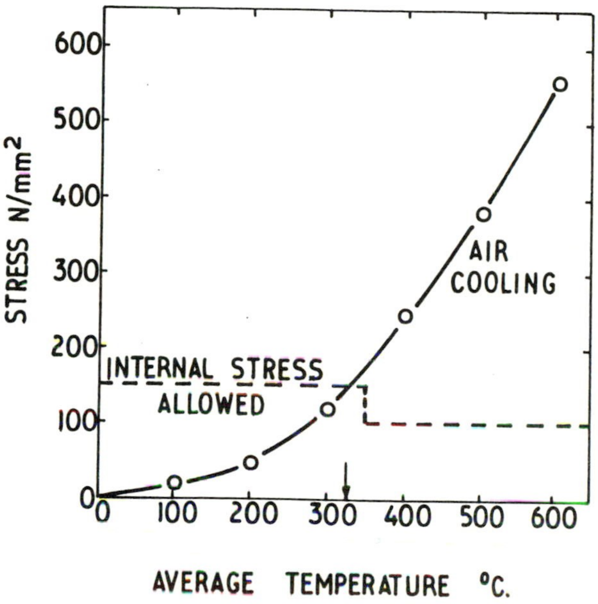

600
500
400
300
200
100
0
STRESS N/mm²
INTERNAL STRESS
ALLOWED
AIR
COOLING
0
100
200
300
400
500
600
AVERAGE TEMPERATURE °C.

Fig. 4.7

SECTION 5 - EXTRUSION

Extrusion is usually carried out hot and results in large deformations, frequently to complex sections. The determination of extrusion pressures can, however, be carried out with reasonable precision from relatively simple equations if the flow stress-strain rate-temperature properties of the material are known and the appropriate temperature can be estimated. Estimation of the temperature conditions during extrusion requires sophisticated computer calculations if the errors introduced in pressure predictions by its uncertainty are not to be greater than those introduced by the use of the relatively simple equations.

5.1 Extrusion Pressure

The pressure required for extrusion depends on the mean flow stress of the material ($\bar{\sigma}$) and on the conditions of extrusion. For backward extrusion or for forward extrusion with zero friction the pressure (P_h) required if deformation were homogeneous would be

$$P_h = \bar{\sigma} \ \ell n \ ER \qquad (5.1)$$

where ER is the extrusion ratio (area of billet (A_o): area of product (A_f)), i.e. the homogeneous strain is

$$\varepsilon_h = \ell n \ ER \qquad (5.2)$$

In practice work must also be done against friction at the die and redundant shear strains are produced so that P_h seriously underestimates real extrusion pressures. In forward extrusion the billet/container friction is also important - this will be considered in example 5.2.

For well lubricated extrusion, such as glass lubricated hot extrusion of steel, the extrusion pressure (P_o) may be expressed as

$$P_o = \alpha \ \bar{\sigma} \ \ell n \ ER \qquad (5.3)$$

where the correction factor α has a value $\simeq$ 1.34 for extrusion through flat dies to round bar (Hughes et al, 1974). More usually the correction for flat dies is given by

$$P_o = \bar{\sigma}(a + b \ \ell n \ ER) \qquad (5.4)$$

where from theory, Kudo (1960) found a = 0.88 and b = 1.30 for smooth dies (zero friction) and a = 1.06 and b = 1.55 for rough dies (sticking friction). In practice the friction conditions have some intermediate value between these limits and intermediate values of the constants, e.g. a = 0.9 and b = 1.5, result (Parkins, 1968).

Reducing the included angle of the die tends to reduce the pressure, but the effect is relatively small in hot extrusion practice and is far outweighed by the effect of the radius of curvature between the die face and die orifice, larger radius leading to higher pressure.

Most commercial extrusions are to complex sections and hollows, and multihole dies may be used. The pressures required are higher than for extrusion to round bar with an equivalent extrusion ratio. An estimate of the pressures required may be obtained by modifying the extrusion ratio in equations (5.3) or (5.4) to

$$ER' \simeq (p_e/p_r)^{\frac{1}{2}}\ ER \qquad (5.5)$$

where p_e is the periphery (external + internal) of the extruded product(s) and p_r is the circumference of round bar having the same extrusion ratio (Sheppard and Wood, 1980).

Example - Determine the value of $P_o/\bar{\sigma}$ for glass lubricated extrusion of steel from 158 mm dia billet to a 40 mm square section with a central hole of 25 mm diameter.

$$A_o = \frac{\pi\ 158^2}{4} = 19607\ mm^2$$

$$A_f = 40^2 - \frac{\pi\ 25^2}{4} = 1109\ mm^2$$

Therefore ER = 17.7

Diameter of round bar of equivalent area

$$D_f = \frac{D_o}{\sqrt{ER}} = \frac{158}{\sqrt{17.7}} = 37.5\ mm$$

Periphery of round bar, $p_r = \pi D_f = 118$ mm

Periphery (external + internal) of section,

$$p_e = 4 \times 40 + \pi \times 50 = 317 \text{ mm}$$

From equation (5.5)

$$ER' \simeq \left(\frac{317}{118}\right)^{\frac{1}{2}} ER = 29.0$$

From equation (5.3)

$$P_0/\bar{\sigma} \simeq 1.34 \ \ell n \ 29.0$$
$$\simeq 4.51$$

Note that for the equivalent round bar

$$P_0/\bar{\sigma} = 1.34 \ \ell n \ 17.7$$
$$= 3.85$$

i.e. the section requires about 17.1% higher pressure, but as the area over which the pressure acts is reduced by the area of the mandrel for the section, the expected extrusion load is only about 16.7% higher than for the equivalent round bar.

5.2 Effect of billet/container friction

In the direct extrusion process, the billet slides through the container and the work done at the interface contributes to the total extrusion pressure (P_e). For sliding friction conditions, analysis of the stresses acting (Rowe, 1965) gives

$$P_e = (P_0 - \bar{\sigma})\exp \frac{4\mu L}{D_o} + \bar{\sigma} \qquad (5.6)$$

where L is the instantaneous length of the billet, measured from the die, and μ is the effective coefficient of friction. For lubricated extrusion, μ is small, e.g. < 0.01 for glass and $\lesssim 0.03$ for graphite based lubricants, and equation (5.6) is frequently simplified to

$$P_e = P_0\left(1 + \frac{4\mu L}{D_o}\right) \qquad (5.7)$$

As μ is usually obtained experimentally by applying equation (5.7)to pressures measured on billets of different initial lengths, any errors in the simplification are masked in the value of μ.

For unlubricated extrusion the shear stress at the interface (or at a skull left in the container) is limited by the shear strength (τ) of the material. In this case the relationship is

$$P_e = P_o + \frac{4\tau L}{D_o} \tag{5.8}$$

and, if equation (5.7) is applied to billets of different lengths, an apparent value of $\mu = \tau/P_o$ is obtained. Note that as P_o is frequently greater than $5\bar{\sigma}$ and taking $\tau \simeq \bar{\sigma}/2$, it can be seen that low apparent values of μ will still be found experimentally.

In the extrusion of hollows using a mandrel of diameter d_o passing through the billet, D_o in equations (5.7) and (5.8) should be replaced by $(D_o - d_o)$

Example

Estimate the load required to extrude an aluminium alloy billet initially 290 mm diameter and 900 mm long in a press with a container of diameter 300 mm, which is heated to the same temperature as the preheat temperature of the billet, if the extruded product is an I section 5 mm thick with flanges 150 mm wide and a web of 200 mm. Assume that the dies are flat and unlubricated and that the appropriate mean flow stress of the material for these specific extrusion conditions is 45 N/mm^2 and that the apparent coefficient of friction between the billet and container is 0.06.

$$A_o = \frac{\pi\ 300^2}{4} = 70686\ mm^2$$

$$A_f = 2 \times 150 \times 5 + 200 \times 5 = 2500\ mm^2$$

Therefore ER = 28.3

Diameter of equivalent round bar = $\left(\frac{4}{\pi} \times 2500\right)^{\frac{1}{2}} = 56.4$ mm

$$p_r = \pi\ 56.4 = 177.2\ mm$$

$$p_e = 2 \times 150 + 4 \times 5 + 4 \times 72.5 + 2 \times 200$$
$$= 1010\ mm$$

From equation (5.5)

$$ER' = \left(\frac{1010}{177.2}\right)^{\frac{1}{2}} 28.3 = 67.6$$

Substitution into equation (5.4) with the appropriate values of the constants gives

$$P_o = 45(0.9 + 1.5 \ln 67.6)$$
$$= \underline{325 \text{ N/mm}^2}$$

The maximum pressure will occur immediately after the billet has been upset in the container, when

$$L = 900 \left(\frac{290}{300}\right)^2 = 841 \text{ mm}$$

From equation (5.7)

$$P_e = 325 \left(1 + \frac{4 \times 0.06 \times 841}{300}\right)$$
$$= \underline{544 \text{ N/mm}^2}$$

Extrusion load = 544 x 70686 = <u>38.5 MN</u>

In practice, steady state conditions of metal flow and temperature are not established immediately after upset and the initial pressure may rise above the calculated values, leading to load requirements for "break-through" of up to about 15% greater than the calculated value. This fractional increment of load varies both with alloy composition and with the extrusion conditions (Sheppard and Wood, 1980; Castle and Sheppard, 1976). Furthermore, because of temperature changes during extrusion (see examples 5.4 and 5.5) the mean flow stress and hence P_o may not be constant throughout the extrusion stroke.

5.3 Mean Strain Rate

During extrusion the strain rate changes continuously through the deformation zone, rising to a maximum just before the material emerges from the die. For the simplest conditions of *homogeneous* deformation of round billet of diameter D_o extruded to round bar of diameter D_f through a die of included semi-angle ϕ the instantaneous strain rate is given by

$$\dot{\varepsilon} = \frac{4\, v_o^2\, D_o \tan \phi}{D^3} \qquad (5.9)$$

where v_o is the ram velocity and D is the instantaneous

value of the diameter (see diagram).

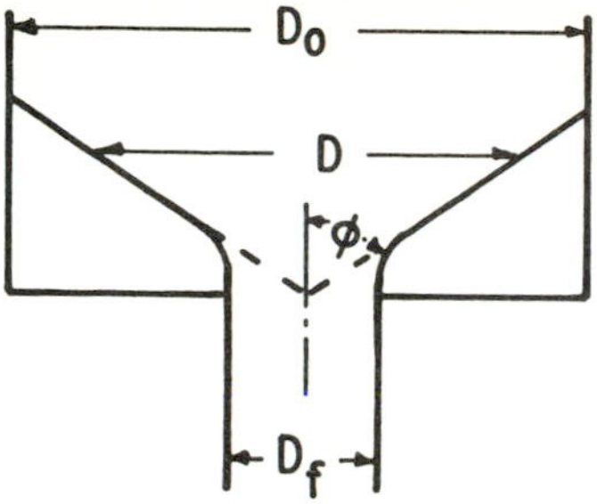

In defining mean strain rate from this equation it is usual to take the time average i.e.

$$\bar{\dot{\varepsilon}}_t = \frac{1}{t}\int_o^t \dot{\varepsilon}\, dt = \frac{\ell n ER}{t} \tag{5.10}$$

where t is the time taken for material to pass through the conical die and ER is the extrusion ratio (homogeneous strain = ℓnER). This leads to the relationship (Wilcox and Whitton, 1958-59)

$$\bar{\dot{\varepsilon}}_t = \frac{6\ v_o\ D_o^2\ \tan\phi}{(D_o^3 - D_f^3)} \cdot \ell n\ ER \tag{5.11}$$

However, when flow stress varies with strain rate and temperature as shown in figure 1.14, i.e.

$$\sigma = \frac{1}{\beta}\left(\ell n\dot{\varepsilon}/_{A'} + \frac{Q}{RT}\right) \tag{5.12}$$

it is more appropriate to take the mean of $\ell n\dot{\varepsilon}$ to obtain the best estimate of $\bar{\sigma}$ for extrusion (Farag and Sellars, 1973). This leads to the relationship

$$\bar{\dot{\varepsilon}}_{ext} = \frac{4\ D_o^2\ v_o\ \tan\phi}{(D_o D_f)^{3/2}} \tag{5.13}$$

It is arguable that both equations (5.11) and (5.13) will underestimate the mean strain rate when redundant strains occur. However, in practice this results in deformation starting before the material enters the conical lead-in, which reduces the mean strain rate. To a first approximation these effects tend to cancel out. On the same basis, extrusion to non-circular sections may be considered to result in the same mean strain rate as extrusion to round bar of equivalent area without excessive error.

In practice flat dies are frequently used. For glass lubricated extrusion the angle ϕ is then determined by flow over the lubricant pad on the die face and is $\sim 70^{o}$. For non-lubricated extrusion ϕ is determined by the angle of the dead-metal zone, and from a minimised upper bound solution (Sheppard, 1981), this is related to the extrusion ratio as

$$\phi = 54.1 + 3.45 \ln ER \quad \text{degrees} \tag{5.14}$$

Example - Determine the mean strain rate for unlubricated extrusion of copper tube shells 100 mm O.D. x 85 mm I.D. from a container of 250 mm diameter using a ram speed of 65 mm/sec.

$$A_o = \frac{\pi}{4} 250^2 = 49087 \text{ mm}^2$$

$$A_f = \frac{\pi}{4}(100^2 - 85^2) = 2179 \text{ mm}^2$$

Therefore ER = 22.5

Equivalent diameter of round bar

$$D_1 = \left(\frac{4}{\pi} 2179\right)^{\frac{1}{2}} = 52.7 \text{ mm}$$

From equation (5.14)

$$\phi = 54.1 + 3.45 \ln 22.5$$
$$= 64.8^{o}$$

Substitution into equation (5.11) gives

$$\dot{\bar{\varepsilon}}_t = \frac{6 \times 65 \times (250)^2 \tan 64.8}{(250^3 - 52.7^3)} \ln 22.5$$

$$= 10.4 \text{ s}^{-1}$$

Substitution into equation (5.13) gives

$$\dot{\bar{\varepsilon}}_{ext} = \frac{4 \times (250)^2 \times 65 \tan 64.8}{(250 \times 52.7)^{3/2}}$$

$$= 22.8 \text{ s}^{-1}$$

Note that strain rates obtained from equation (5.13) are

always higher than those obtained from equation (5.11) and hence when used to determine mean flow stress (see example 5.5) they always lead to somewhat higher values.

5.4 Temperature Rise during Extrusion

During hot extrusion, virtually all the work done is converted to heat, leading to a temperature rise in the material. Three contributions arise from the work:

(a) of upsetting

(b) against billet/container friction

(c) of extrusion through the die.

Contribution (a) is small and is frequently neglected, but contributions (b) and (c) must not be overlooked. The temperature rise from them is not uniform across the section, but can be expressed in terms of a rise in *mean* temperature ($\bar{T}$) of a cross section, defined so that the heat content of the section with a uniform temperature of $\bar{T}$ would be the same as with the existing temperature gradient, i.e.

$$\bar{T} = \frac{8}{D^2} \int_o^{D/2} r \, T_r \, dr \qquad (5.15)$$

The rise in mean temperature due to billet/container friction ($\Delta\bar{T}_f$) can then simply be calculated from the change in pressure with distance. If it is assumed that half the work done appears as heat in the billet and half as heat in the container,

$$\Delta\bar{T}_f = \frac{1}{2s\rho} (P_e - P_o) \qquad (5.16)$$

where P_e is the initial pressure on a section initially at distance ℓ above the die, P_o is as defined previously and s and ρ are the specific heat and density.

Similarly the rise in mean temperature due to extrusion through the die is

$$\Delta\bar{T}_{ext} = \frac{P_o}{s\rho} \qquad (5.17)$$

<u>Example</u>

For the conditions given in example 5.2 calculate the temperature rise due to friction and deformation at the front end of the extrusion and at the back end of the extrusion, assuming that 20 mm of the billet remains in the container as the discard.

Specific heat = 1080 J/kgK, Density = 2750 kg/m^3

From example 5.2, the pressure at the deformation zone P_o = 325 N/mm^2 throughout the extrusion.

Substituting into equation (5.17) gives the temperature rise during extrusion.

$$\Delta\bar{T}_{ext} = \frac{325 \times 10^6}{1080 \times 2750}$$

$$= 109^{o}C$$

At the front end there is no additional temperature rise due to friction between the billet and container, but 20 mm below the pressure pad the section of billet slides a distance

$$\ell = 841 - 20 = 821 \text{ mm}$$

Initially the pressure P_e at this section, from equation (5.7), is

$$P_e = 325 \left(1 + \frac{4 \times 0.06 \times 821}{300}\right)$$

$$= 538 \text{ N/mm}^2$$

Therefore, from equation (5.16)

$$\Delta\bar{T}_f = \frac{(538 - 325) \times 10^6}{2 \times 1080 \times 2750}$$

$$= 36^{o}C$$

At the front end the total temperature rise = 109°C

At the back end the total temperature rise = 145°C

Note the magnitude of these temperature rises. It is important that they are taken into account in selecting the conditions for extrusion (see example 5.6).

In practice there are also heat losses during extrusion that must be taken into account in calculating extrusion temperatures. These arise from:

(i) convection and radiation to the environment during transfer from the reheating unit to the container

(ii) conduction to the container

(iii) conduction to the die (and lubricant pad)

(iv) conduction from the deformation zone to the undeformed part of the billet.

In the extrusion of aluminium alloys the temperature is relatively low and the container may be at nearly the same temperature so that (i) and (ii) may be small, but as ram speeds are relatively low, e.g. 10 mm/sec, (iii) and (iv) are important. In contrast, in the extrusion of steel and other high temperature alloys (i) and (ii) are important, but because ram speeds are generally higher, e.g. 75 mm/sec, (iii) and (iv) may be small. The complexity of these heat losses means that computer methods must be applied to obtain reasonable estimates of the temperature changes (Sheppard and Wood, 1980; Sellars and Whiteman, 1981).

5.5 Change in Mean Flow Stress and Extrusion Pressure with Ram Travel

At the high strains involved in hot extrusion the flow stress of the material has usually reached a steady state value and is no longer sensitive to strain, but, as shown in example 1.5, it is sensitive to strain rate and temperature and in the range of interest may follow a relationship

$$Z = \dot{\varepsilon} \exp \frac{Q}{RT} = A'_{ss} \exp(\beta\, \sigma_{ss}) \tag{5.18}$$

where for extrusion conditions $\dot{\varepsilon}$ is replaced by $\dot{\bar{\varepsilon}}_{ext}$ (or $\dot{\bar{\varepsilon}}_t$) (see example 5.3), and T is replaced by $\bar{T}_{ext}$, the mean extrusion temperature, which can be defined as

$$\frac{1}{\bar{T}_{ext}} \simeq \frac{1}{2}\left(\frac{1}{\bar{T}_o} + \frac{1}{\bar{T}_e}\right) \tag{5.19}$$

where $\bar{T}_o$ is the mean temperature of a section of billet at entry to the deformation zone and $\bar{T}_e$ is the mean temperature of the section on emergence from the die.

For extrusion of steels or high temperature alloys

$$\bar{T}_{ext} \simeq \bar{T}_o + \frac{1}{2}\Delta\bar{T}_{ext} \qquad (5.20)$$

where $\Delta\bar{T}_{ext}$ is defined by equation (5.17).

Calculation of mean flow stress and thence extrusion pressure from basic flow stress data requires a knowledge of the heat flow affecting $\bar{T}_o$ as well as of the extrusion conditions and an iteration must therefore be made to find $\Delta\bar{T}_{ext}$.

Example

Determine the mean flow stress and the extrusion pressure at the beginning of steady state extrusion (40 mm ram travel after upset) and at the end of extrusion (25 mm discard) for the extrusion in example (5.1) if the billet of initial length 450 mm and diameter 158 mm is reheated to 1150°C, the ram speed is 60 mm/sec, the effective coefficient of billet/container friction is 0.008 and the material is mild steel with the properties given in example (1.5), i.e.

$$Z = \dot{\varepsilon} \exp \frac{305\ 000}{8.31\ T} = 1.03 \times 10^9 \exp 7.67 \times 10^{-2}\sigma_{ss}$$

when $Z > 10^{11} s^{-1}$

A reheating temperature of 1150°C gives $\bar{T}_o$ at 40 mm = 1128°C and at the end = 1088°C (estimated from the master curves derived from finite difference computations as described by Sellars and Whiteman (1981) for a container temperature of 300°C). The product $s\rho$ for mild steel in the austenitic condition is 5×10^6 J/m^{3}K over the temperature range of interest.

This calculation requires iteration and is carried out in the following steps:-

Calculate $\dot{\bar{\varepsilon}}_{ext}$

Assume $\Delta\bar{T}_{ext} = 0$ (or some other value)

Calculate Z

Determine σ_{ss}

Calculate P_o

Calculate $\Delta\bar{T}_{ext}$

(iterate from "Calculate Z" to "Calculate $\Delta\bar{T}_{ext}$" until there is no significant change in the value of $\Delta\bar{T}_{ext}$)

Calculate P_e

From equation (5.13), taking $\phi = 70^{\circ}$

$$\dot{\bar{\varepsilon}}_{ext} = \frac{4 \times 158^2 \times 60 \times \tan 70}{(158 \times 37.5)^{3/2}}$$

$$= 36.1 \text{ s}^{-1}$$

For 40 mm ram travel, initially take

$$\bar{T}_{ext} = \bar{T}_o = 1128^{\circ}\text{C} = 1401 \text{ K}$$

Then $Z = 36.1 \exp \dfrac{305\ 000}{8.31 \times 1401} = 8.61 \times 10^{12} \text{ s}^{-1}$

and $\sigma_{ss} = \dfrac{1}{7.67 \times 10^{-2}} \ln \dfrac{8.61 \times 10^{12}}{1.03 \times 10^{9}}$

$$= 118 \text{ N/mm}^2$$

From example (5.1)

$$P_o = 4.51\ \bar{\sigma} = 531 \text{ N/mm}^2$$

From equation (5.17)

$$\Delta\bar{T}_{ext} = \frac{531 \times 10^6}{5 \times 10^6} = \underline{106^{\circ}\text{C}}$$

Thus, from equation (5.20) the next estimate of

$$\bar{T}_{ext} = 1128 + 53 = 1181^{\circ}\text{C} = 1454 \text{ K}$$

and proceeding as above gives the next estimate of

$$\Delta\bar{T}_{ext} = \underline{95^{\circ}\text{C}}$$

Thus the next estimate of

$$\bar{T}_{ext} = 1128 + 48 = 1176^{\circ}\text{C} = 1449 \text{ K}$$

Taking this value as a reasonable estimate of $\bar{T}_{ext}$ gives

$$\bar{\sigma} = 106 \text{ N/mm}^2$$

$$P_o = 480 \text{ N/mm}^2$$

and $\Delta\bar{T}_{ext} = 96^\circ C$ (i.e. negligibly different from the previous estimate)

To find the extrusion pressure, the length of billet remaining after 40 mm ram travel must first be found as

$$\ell = 450 \times \left(\frac{150}{158}\right)^2 - 40$$

$$= 406 - 40 = 366 \text{ mm}$$

Then applying equation (5.7)

$$P_e = 480\left(1 + \frac{4 \times 0.008 \times 366}{158}\right)$$

$$= 516 \text{ N/mm}^2$$

For the end of extrusion, initially take $\Delta\bar{T}_{ext} = 104^\circ C$, from previous calculation, then, proceeding as before

$$\bar{T}_{ext} = 1088 + 52 = 1140^\circ C = 1413 \text{ K}$$

$$\bar{\sigma} = 115 \text{ N/mm}^2$$

$$P_o = 518 \text{ N/mm}^2$$

and $\Delta\bar{T}_{ext} = 104^\circ C$, so no iteration is required.

Applying equation (5.7)

$$P_e = 518\left(1 + \frac{4 \times 0.008 \times 25}{158}\right)$$

$$= 521 \text{ N/mm}^2$$

Note that the effect of temperature on σ has a greater influence on P_e than the small coefficient of friction, leading to a slightly rising pressure with ram travel.

5.6 Extrusion Limit Diagram

The extrusion ratio attainable by hot extrusion is generally limited by one of two factors: (1) excessive pressure requirement (2) excessive temperature rise leading to a defective product. These limitations may be shown as a function of reheating temperature to give an extrusion limit diagram for satisfactory extrusion (Hirst and Ursell, 1958).

In the case of glass lubricated extrusion a lower temperature limit is also imposed by the softening temperature of the glass.

Because of the temperature sensitivity of the properties of materials, a diagram must be obtained for a particular billet size, ram speed and container temperature.

Example

Determine the extrusion limit diagram for glass lubricated extrusion of type 316 stainless steel to round bar from billets 125 mm diameter x 250 mm long after upset in a vertical press with a container temperature of $300^{\circ}C$ and a ram speed of 75 mm/s^{-1}. A discard of 20 mm is to be left in the container.

Assume that α in equation (5.3) is 1.34, μ in equation(5.7) is 0.01, ϕ in equation (5.13) is 70° and the maximum allowable pressure is 1300 N/mm^2. The steady state flow stress for type 316 stainless steel is (Nair et al, 1974)

$$Z = \dot{\varepsilon} \exp \frac{460\ 000}{8.31\ T} = 1.48 \times 10^{14} \exp 5.60 \times 10^{-2} \sigma_{ss}$$

and the steel has satisfactory ductility for extrusion to a maximum temperature of $1400^{\circ}C$. Burning and unsatisfactory extrusion may be assumed to occur if the *mean* temperature exceeds this value at any stage of the process. The product $s\rho$ for stainless steel is 5×10^6 J/m^3K over the temperature range of interest.

Because the press is vertical, additional chilling of the front end of the billet takes place by longitudinal heat flow to the lubricant pad before upset. From information about this and from the master curves derived from finite difference computations for radial heat flow (Sellars and Whiteman, 1981), the relationships between reheating temperature and (a) the mean temperature at the peak pressure (which occurs at 5 mm ram travel after upset), (b) at 35 mm ram travel after upset (when the additional chilling is negligible, i.e. the mean temperature is a maximum) and (c) at the end of extrusion are shown in figure 5.1. Similar calculations for surface temperature at the end of extrusion also show that the minimum reheating temperature for satisfactory lubrication is 980°C.

In the extrusion limit diagram, figure 5.2, the vertical lines at 980°C, the lower limit for satisfactory extrusion, and at 1400°C, above which burning during reheating will occur, may be drawn.

<u>Pressure limitation</u> - because the minimum value of $\bar{T}_o$ occurs at the initial peak pressure, this will be the maximum pressure in the extrusion curve. The limiting ER is therefore determined for the peak, when L = 245 mm.

From equation (5.7)

$$P_e = P_o\left(1 + \frac{4 \times 0.01 \times 245}{125}\right)$$
$$= 1.078 \, P_o$$

Therefore when P_e = 1300 N/mm², P_o = 1205 N/mm².

As the extrusion ratio affects the strain rate, to avoid iteration it is necessary to assume values of limiting ER and calculate the appropriate temperature. First, try ER = 100 and then appropriate multiples when the result for this value is known: see table below.

(1) ER	(2) D_1 mm	(3) $\dot{\varepsilon}$ s^{-1}	(4) $\bar{\sigma}$ N/mm²	(5) Z s^{-1}	(6) $\bar{T}_{ext}$ °C	(7) $\bar{T}_o$ °C	(8) RHT °C
100	12.50	209	195	8.2×10^{18}	1176	1055	1172
200	8.84	351	170	2.0×10^{18}	1252	1131	1292
60	16.14	142	220	3.3×10^{19}	1111	990	1081
40	19.76	105	244	1.3×10^{20}	1056	935	1000

Method of calculation:

Column (2) from (1) as ER = D_o^2/D_f^2

(3) from (2) and equation (5.13) with v_o = 75 mm s^{-1}

(4) from (1), equation (5.3) and the value of $P_{o(max)}$

(5) from (4) and the equation for steady state flow stress

(6) from (5) and (3) as $\bar{T}_{ext}(K) = Q/R\ \ell n(Z/\dot{\varepsilon})$

(7) from (6) and equations (5.17) and (5.20)

(8) from (7) and figure 5.1

From columns (1) and (8) line (1) is plotted in figure 5.2.

Burning during extrusion - to avoid this it is required that

$$\bar{T}_{o(max)} + \Delta\bar{T}_{ext} \leq 1400^{o}C$$

where $\bar{T}_{o(max)}$ occurs at 35 mm ram travel after upset. As temperature is now specified, the maximum value of ER must be calculated, which involves iteration because of the effect of ER on strain rate. Select a value of $\Delta\bar{T}_{ext}$ and then choose other values when the results from this value is known, see table below:-

(1)	(2)	(3)	(4)	(5)	(6)
$\Delta\bar{T}_{ext}$ ^{o}C	$\bar{T}_o$ ^{o}C	RHT ^{o}C	$\bar{T}_{ext}$ ^{o}C	P_o N/mm^2	ER
150	1250	1292	1325	750	78
200	1200	1248	1300	1000	158
100	1300	1348	1350	500	30
0	1400	~1460	1400	0	1

Method of calculation:

Column (2) from (1) and the criterion for burning

(3) from (2) and figure 5.1 for 35 mm

(4) from (1) and equation (5.20)

(5) from (1) and equation (5.17)

(6) from (5) and equation (5.3) with iteration to find ER using the equation for steady state flow stress, e.g.

when $\Delta\bar{T}_{ext}$ = 150°C, $\bar{\sigma}\ \ell n$ ER = 560 N/mm^2, then guess a value of ER, say 60.

Then from ER = D_o^2/D_f^2 and equation (5.13) with v_o = 75 mm s^{-1} $\dot{\varepsilon}$ = 142 s^{-1}.

Hence, with $\bar{T}_{ext}$ from column (4) $Z = 1.6 \times 10^{17}\ s^{-1}$
from the equation for steady state stress $\bar{\sigma} = 124\ N/mm^2$.

Therefore $ER = \exp \frac{560}{124} = 91$

This gives $\dot{\varepsilon} = 195\ s^{-1}$ and proceeding as before gives $\bar{\sigma} = 130\ N/mm^2$ and ER = 75.

Repeating gives ER = 81, then ER = 78, although these last two iterations are not really required within the precision of the prediction.

From columns (3) and (6) line (2) is plotted in figure 5.2.

Note that because of cooling after reheating, a value of RHT > 1400°C is predicted for low values of $\Delta\bar{T}_{ext}$. In practice burning would have occurred during reheating so the upper temperature limit would be as estimated by the chain line in figure 5.2. The area within the lines now defines the limiting conditions for satisfactory extrusion.

If extrusion to sections other than rounds is to be done, ER in the diagram must be replaced by ER′ as defined by equation (5.5) so the real limiting values of extrusion ratio will be less than shown in the figure by an amount that can be determined from equation (5.5).

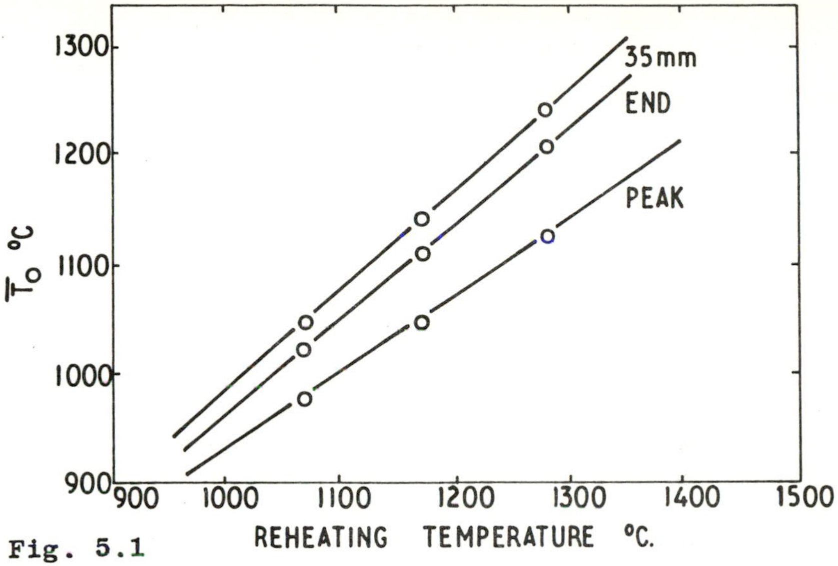
1300
1200
1100
1000
900
$\overline{T}_o$ °C
35mm
END
PEAK
900
1000
1100
1200
1300
1400
1500
REHEATING TEMPERATURE °C.

Fig. 5.1

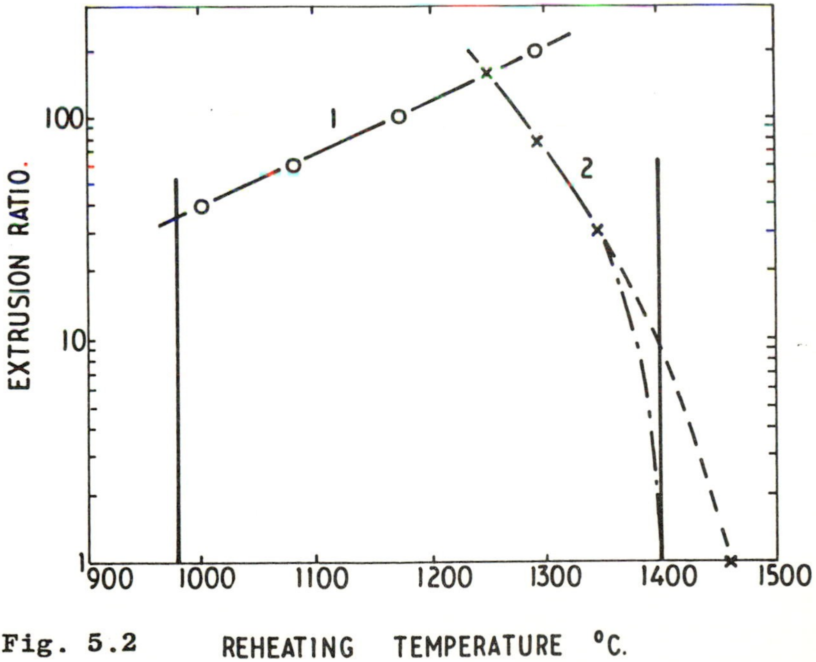
100
10
1
EXTRUSION RATIO.
1
2
900
1000
1100
1200
1300
1400
1500
REHEATING TEMPERATURE °C.

Fig. 5.2

SECTION 6 - WIRE DRAWING

Wire drawing appears to be one of the simplest of the working processes, but in some ways it is one of the more complicated to analyse. The effects of friction and redundant work contribute a major part of the power requirements for drawing and in many cases the work expended in overcoming these two factors can exceed that for homogeneous deformation. It is therefore important to minimise these effects so as to achieve the most economical drawing conditions. As friction and redundant work are influenced in opposite ways by die angle, a compromise has to be made to obtain optimum conditions.

A secondary effect of redundant work is to cause non-homogeneous deformation which results in wire with steep property gradients between the centre-line and the surface. This is unacceptable for many applications. A knowledge of how to minimise the effects of friction and redundant work can thus lead to the production of wire with more uniform properties.

Numerous equations have been proposed for the prediction of drawing loads and it is difficult to select a single one as being the most satisfactory. For this reason several of the more widely accepted equations have been used in this section to illustrate the essential similarities between them. Considerable attention has also been given to the treatment of non-homogeneous deformation since much of the literature concerning this is confusing. Finally examples are given relating to multiple-pass sequences and to the scheduling of multi-hole commercial wire drawing.

6.1 Drawing Stress

In wire drawing the strain in the material is given by the elongation produced in a single pass, or assuming constant volume, by the area before and after passing through the die.

$$\varepsilon_w = \ell n \frac{\ell_f}{\ell_o} = \ell n \frac{A_o}{A_f}$$

Now, if we define the fractional reduction in area as

$$r = \frac{A_o - A_f}{A_o} \tag{6.1}$$

and therefore r is *positive*, even though the deformation is compressive, the strain can be written

$$\varepsilon_w = \ln \frac{1}{1-r} \tag{6.2}$$

The drawing stress σ_d may be obtained from a consideration of the work done by the force F in drawing the wire, and including a factor β to account for both redundant work and friction

$$\sigma_d = \frac{F}{A_f} = \beta\, \bar{\sigma} \ln \frac{1}{1-r} \tag{6.3}$$

where $\bar{\sigma}$ is the mean flow stress of the material over the appropriate strain interval.

This equation is adequate for many order-of-magnitude calculations with β having a numerical value between about one and three. If the factor β is given the value unity then the equation gives the stress required to deform the wire homogeneously to a reduction r. β is thus the ratio of the total work to the homogeneous work.

The factor β is sometimes expressed in the form of an efficiency factor η, where $\eta = \frac{1}{\beta}$. Similarly η is the product of two component efficiencies η_ϕ and η_f, the efficiency due to redundant work losses, and the efficiency due to frictional losses, respectively (see later). η is by definition less than unity.

The load F is obtained from (6.3) by multiplying the stress by the exit area.

<u>Example</u> - Find the drawing load required to reduce copper wire by 36% reduction in area in a single pass to 2 mm diameter, if the mean flow stress in the pass is 250 N/mm^2, and 50% of the total load is used in overcoming friction and redundant work.

Since 50% of the load is used in overcoming friction and redundant work then the other 50% is used in drawing homogeneously.

Thus if $\beta = 1$ represents homogeneous deformation, then $\beta = 2.0$ represents the total deformation.
For 36% Reduction $r = 0.36$
Thus in equation (6.3) $F = \frac{\pi(2)^2}{4} \times 2.0 \times 250 \; \ell n \left(\frac{1}{1 - 0.36} \right)$

$$F = \underline{701 \text{ N} = 71.5 \text{ kgf}}$$

6.2 Drawing Stress involving Friction and Redundant Work

There are many other expressions published in the literature that are more rigorous than equation (6.3) and that take into account the die geometry, the coefficient of friction and redundant work. One equation that Wistreich (1958) feels is preferable to many others is given by Siebel (1947).

$$\sigma_d = \frac{F}{A_f} = \bar{\sigma}\left\{\left(1 + \frac{\mu}{\alpha}\right)\ell n \frac{A_o}{A_f} + \frac{2\alpha}{3}\right\} \qquad (6.4)$$

where μ is the friction coefficient, and α is the angle of taper of the die (often called the die semi-angle). The last term in the expression takes account of redundant deformation. This equation is favoured for its simplicity and because although not exact, the underlying premises are reasonably realistic.

Another equation that is much more empirical in its treatment of redundant work is that due to Whitton (1957-58) who extended the earlier equation of Sachs (1927), (which neglected the effects of redundant work). The equation has been shown to give results within ± 10% over most of the usual wire drawing ranges, but it is not satisfactory for reductions of less than 10% or possibly for reductions greater than 50%, Wistreich (1958).

$$\sigma_d = \frac{F}{A_f} = \bar{\sigma}\left(1 + \frac{1}{\mu\cot\alpha}\right)\left\{1 - \left(\frac{A_f}{A_o}\right)^{\mu\cot\alpha}\right\} + \frac{2}{3}\bar{\sigma}\,\alpha^2\left(\frac{1-r}{r}\right) \qquad (6.5a)$$

$$= \bar{\sigma}\left(\frac{1+B}{B}\right)\left\{1 - (1-r)^B\right\} + \frac{2}{3}\bar{\sigma}\,\alpha^2\left(\frac{1-r}{r}\right) \qquad (6.5b)$$

where $B = \mu\cot\alpha$

The first term in equation (6.5) accounts for the presence of friction whilst again, the last term accounts for redundant work.

Whitton has additionally pointed out the importance of using the correct value of $\bar{\sigma}$ if these equations are to be used for an estimate of μ, since any small error in $\bar{\sigma}$ is magnified by over an order of magnitude in the determination of μ.

Example - In the previous problem, if the average coefficient of friction for drawn copper is 0.1 and the angle of taper of a conical die is 14°, determine the drawing loads using the methods of both Siebel and Whitton.

$$\text{Now } 14^{o} = 0.2444 \text{ rad and } \frac{A_f}{A_o} = 1 - 0.36 = 0.64$$

Siebel : equation (6.4)

$$F = \frac{\pi(2)^2}{4} \times 250 \left\{ \left(1 + \frac{0.1}{0.2444}\right) \ell n \frac{1}{0.64} + \frac{2 \times 0.2444}{3} \right\}$$

$$= \pi \times 250 \{0.4463 + 0.1826 + 0.1629\}$$

$$F = \underline{621.9 \text{ N}} = \underline{63.4 \text{ kgf}}$$

Note that the three numbers in the bracket above represent the proportions of work expended due to homogeneous deformation, friction and redundant work, respectively.

Whitton : equation (6.5a)

$$\mu\cot\alpha = \mu\tan(90 - \alpha) = 0.1 \tan 76^{o}$$
$$= 0.4011$$

$$F = \frac{\pi(2)^2}{4} \times 250 \left\{ \left(1 + \frac{1}{0.4011}\right) \left[1 - (0.64)^{0.4011}\right] + \frac{2}{3} (0.2444)^2 \frac{0.64}{0.36} \right\}$$

$$= \pi\ 250 \left\{(3.493)(0.1639) + 0.0708\right\}$$

$$F = \underline{505.2 \text{ N}} = \underline{51.5 \text{ kgf}}$$

The equations (6.4 and 6.5) deal with redundant work by adding an extra term to the basic equations. There is, however, some merit in dealing with redundant work by using a multiplying factor on the basic equations for drawing stress in the presence of friction. The two equations most widely quoted are shown below:

$$\sigma_d = \bar{\sigma}\ (1 + B)\phi_1\ \ln \frac{A_o}{A_f} \qquad (6.6)$$

and

$$\sigma_d = \bar{\sigma}\ \frac{(1 + B)}{B}\phi_1 \left\{1 - \left(\frac{A_f}{A_o}\right)^B\right\} \qquad (6.7)$$

These equations closely resemble the earlier ones.

The term $(1 + B)\phi_1$ in equation (6.6) has the same meaning as β in equation (6.3). Thus $(1 + B), \left(= \frac{1}{\eta_f}\right)$, represents the drawing stress increase due to friction, whilst $\phi_1 = \left(\frac{1}{\eta_\phi}\right)$ represents the increase in stress due to redundant work and is therefore called the *redundant work factor*. Equation (6.6) is derived from an extension of the theory for plane strain strip drawing [Hill and Tupper 1948].

Equation (6.7) is based on the equation by Sachs (1927) and the similarity to (6.5) will be noted. It has been pointed out that when $(B.\varepsilon)$ becomes small then equation (6.7) approaches the value given by equation (6.6) (Atkins and Caddell 1968).

It is found that ϕ_1 is related to the geometry of drawing through a linear equation of the form

$$\phi_1 = k_1 + k_2\ \Delta \qquad (6.8)$$

where k_1 and k_2 are constants and the parameter Δ is given by

$$\Delta = \left(\frac{D_o + D_f}{D_o - D_f}\right) \sin \alpha \approx \frac{4\alpha}{\varepsilon_w} \qquad (6.9)$$

There are several different ways of defining Δ quoted in the literature and these yield different values for a fixed geometry of die. The various definitions are compared and discussed in the Appendix to a paper by Caddell and Atkins (1968). It is therefore important that in using equations containing the factor Δ, its definition in the context of that equation is appreciated.

Equation (6.8) is satisfactory up to values of $\Delta = 6$ after which there is an increasing deviation of the $\phi_1 - \Delta$ curve from linearity. Values of k_1 and k_2 have been determined experimentally on four different metals spanning a wide range of work hardening coefficients, both in the annealed and in the 25% strain hardened condition (Atkins and Caddell 1968). The values are tabulated below:-

Annealed

	k_1	k_2
Copper	1.0	0.27
Brass	1.0	0.27
Mild Steel	0.8	0.09
Aluminium	0.8	0.09

25% Strain Hardened

	k_1	k_2
Copper, Brass, Mild Steel & Aluminium	0.88	0.19 to 0.22

Methods of finding ϕ_1 directly by comparing flow stresses of drawn wires with flow stresses of annealed wires are open to question due to the presence of Bauschinger effects and of residual stresses (Trozera 1964). They will not, therefore, be considered here.

As in previous examples the equations (6.6) and (6.7) both incorporate a mean yield stress rather than use a rigorous expression to describe the work hardening behaviour of the metal. The effect of this on the drawing stress has been investigated by Atkins and Caddell over a range of drawing conditions and work hardening coefficients. They concluded that the drawing stress is underestimated if the mean yield is used but even in the worst cases the error is less than 10% and diminishes with prior cold work. They suggested, therefore, that the extra complexity of using an exact expression for the work hardening behaviour was not, in general, justified.

Example - Annealed Copper wire, diameter 2.5 mm, is drawn through a conical die of taper angle 14°, to a diameter of 2.0 mm. If the frictional coefficient is 0.1, determine the drawing force for a mean flow stress of 250 N/mm².

From equation (6.9)

$$\Delta = \left(\frac{2.5 + 2.0}{2.5 - 2.0}\right) \sin 14^{\circ}$$

$$\Delta = 2.177$$

Thus for annealed copper equation (6.8) gives

$$\phi_1 = 1.0 + 0.27 \times 2.177 = 1.59$$

Hence from equation (6.6):

$$F = A_f \sigma_d = \frac{\pi (2)^2}{4} 250 \, (1 + 0.1 \tan 76^{\circ}) \, 1.59 \, \ln\left(\frac{1}{0.64}\right)$$

$$F = \underline{780.8 \text{ N}} = \underline{79.6 \text{ kgf}}$$

While from equation (6.7):

$$\text{Since } B = 0.1 \tan 76^{\circ} = 0.401$$

$$F = A_f \sigma_d = \frac{\pi (2)^2}{4} 250 \left(\frac{1+0.401}{0.401}\right) \times 1.59 \times \left\{1 - \left(\frac{2}{2.5}\right)^{2 \times 0.401}\right\}$$

$$F = \underline{715.1 \text{ N}} = \underline{72.9 \text{ kgf}}$$

6.3 Optimum Die Angle for Minimum Work

In view of the fact that the efficiency is the product of two terms that act virtually independently and that the redundant work increases as the die semi-angle increases whilst friction decreases, then there is an angular range of taper of the dies α^* where the work required to carry out the deformation is a minimum. It is clearly more economical to operate drawing processes at this minimum. The optimum semi-angle has been shown by Wistreich (1955), to be given approximately by

$$\alpha^* \approx \sqrt{\frac{\mu r}{1-r}} \qquad (6.10)$$

This is a very simple but very useful equation to have to hand as a first estimate in a drawing process. It will be noteu that α^* does not depend on the precise work hardening behaviour of the metal or even on the type of metal being drawn. From this point of view it is particularly useful.

If drawing is carried out at angles greater than the optimum then the load required would increase significantly. Eventually if the angle were much greater than α^* the load might be such that it might be more favourable for the metal to shear at some smaller angle and to form a dead metal zone in the wire in exactly the same way as in extrusion, (section 5), (Backofen 1972).

The various expressions for α^* considered by Wistreich (1955) all contain the coefficient of friction and the reduction in area r within the square root. This means that the confidence that can be placed in α^* is not much better than about ± 5% since the confidence in μ is not much better than ± 10% (see later). However, the load v. die semi-angle curve from which we require the minimum is fairly shallow and so an error in α^* of this magnitude would not be critical.

Example - For the metal and drawing conditions in the previous example, determine the optimum angle of drawing for minimum power requirements.

From equation (6.1)

$$r = 1 - \left(\frac{D_f}{D_o}\right)^2 = 1 - \left(\frac{2.0}{2.5}\right)^2$$

$$r = 0.36$$

In equation (6.10)

$$\alpha^* = \left\{\frac{0.1 \times 0.36}{1 - 0.36}\right\}^{\frac{1}{2}}$$

$$\alpha^* = 0.237 \text{ rad}$$

$$\alpha^* = \underline{13.6^{o}}$$

6.4 Axial Tensile Stress

According to the theory of Siebel (1947) there is a variation in longitudinal stress across the section of the wire during drawing through a conical die. At the wire surface the stress is less than the value of the average drawing stress, whilst along the centre line of the wire it is greater than the applied stress. The excess stress on the centreline above that required for drawing is given by the difference between the centreline stress, σ_{max}, and the average drawing stress. It can thus be shown that:-

$$\sigma_{max} - \frac{F}{A_f} = \frac{2\,\alpha\,\bar{\sigma}}{3(1 + \frac{\mu}{\alpha})} \quad \frac{1}{\ell n\,\frac{A_o}{A_f}} \qquad (6.11)$$

Example (a) - Using the data given in the example in section 6.2 and Siebel's equation (6.4) find the value of the centreline stress

---

Now $$\frac{F}{A_f} = \frac{621.9}{\pi} = 198.0 \text{ N/mm}^2$$

Hence $$\sigma_{max} = 198.0 + \frac{2 \times 0.2444 \times 250}{3\left(1 + \frac{0.1}{0.2444}\right)\ell n\left(\frac{1}{0.64}\right)}$$

$$= 198.0 + 64.8$$

$$\sigma_{max} = \underline{262.8 \text{ N/mm}^2}$$

Example (b) - Centreline fracture of drawn wire is encouraged by both large die angles and by small reductions. Thus, keeping the same angle as above, but with a reduction in area of only 10%, find the centreline stress for these conditions.

In equation (6.4) $$\frac{F}{A_f} = 250\left\{\left(1 + \frac{0.1}{0.2444}\right)\ell n\,\frac{1}{0.9} + \frac{2\text{x}0.2444}{3}\right\}$$

since $$A_f = 0.9\ A_o$$

$$\frac{F}{A_f} = 250\ \{0.1485 + 0.1629\}$$

$$\underline{\text{or}} \quad \frac{F}{A_f} = 77.9\ \text{N/mm}^2$$

In equation (6.11)

$$\sigma_{max} = 77.9 + \frac{2 \times 0.2444 \times 250}{3\left(1 + \frac{0.1}{0.2444}\right)\ell n\ \frac{1}{0.9}}$$

$$= 77.9 + 274.3$$

$$\therefore \quad \sigma_{max} = \underline{352.2\ \text{N/mm}^2}$$

This value of σ_{max} is significantly greater than the average flow stress of the metal. These differential stresses across the section can lead to chevron cracking at the centre of the wire, through the effect on the strain fields within the wire. The condition for this type of cracking has been examined quantitatively by Avitzur (1968), and the range of drawing variables where central cracking might be expected has been defined. In addition he points out that prior annealing of the wire and rounding of the corner between the conical and cylindrical portion of the die will reduce the likelihood of cracking. Although drawing may be within the range where central bursting might be expected, it does not necessarily occur (Avitzur 1970). The bursting phenomenon occurs when the plastic zone fails to penetrate to the centre of the wire and therefore those factors that localise the deformation at the surface might be expected to encourage cracking, eg. small reduction, large die angle and work hardened metal.

6.5 Backpull and Coefficient of Friction

When a backward force Q is applied to the ingoing wire there is a reduction in the axial component of the force on the die, F_D, and the load needed to make the drawing reduction increases.

This can be expressed as

$$F = F_D + Q \tag{6.12}$$

where F is the drawing load.

It can be shown experimentally that there is a linear relation between F_D and Q such that as backpull increases the die load decreases.

$$F_D = F_{D_o} - bQ \tag{6.13}$$

where F_{D_o} $(= F_o)$ is the value of F_D when there is zero retarding force and b, the "back-pull" factor is the gradient of the F_D v. Q curve, $\left(-\dfrac{dF_D}{dQ}\right)$.

One expression for b is given by (Hill 1951)

$$b = (1 + \mu\cot\alpha)r \tag{6.14}$$

This suggests immediately a method of finding the coefficient of friction for wire drawing, without the necessity of knowing anything about the tensile properties of the metal being drawn. It is only necessary to measure the die force and simultaneously the value of the backpull and to plot these on a curve. The average gradient will give an average value of b to substitute into equation (6.14). Since the coefficient of friction is sensitive to both drawing speed and to temperature, care must be taken in controlling the experimental conditions. In addition the coefficient increases slightly as the interfacial pressure between the wire and the die increases, and this leads to slight curvature in the die load-backpull curve. Under most drawing conditions, however, the curvature is negligible (Baron and Thompson 1951). It is nevertheless clear that the values of coefficient measured depend critically on the precise experimental conditions and this sets a limit on the confidence that can be placed on the results.

Example - A 1.5 mm diameter 65/35 brass wire is pulled at 1.08 m/min through a conical die of semi-angle 10.5^{o} with liquid lubricant. For a fixed drawing reduction of 0.37, various retarding forces were applied to the ingoing wire, and the corresponding loads on the die were read. The data are laid out in Table 6.1. Determine the average coefficient of friction for these drawing conditions.

Table 6.1

Back Pull Q (N)	50	150	250	350	450	550
Die Load F_D (N)	380	306	242	185	132	80

The data are plotted in figure 6.1. The conditions have been chosen to show the curvature of the line, but this is greater than many other test conditions.

The average value of $b\left(= -\frac{dF_D}{dQ}\right)$ over the whole range is 0.6.

Thus in equation (6.14) with r = 0.37.

$$\mu = \left(\frac{b}{r} - 1\right)/\cot\alpha$$

$$= \left(\frac{0.6}{0.37} - 1\right)/\tan 79.5^{o}$$

$$\text{or} \quad \mu = \underline{0.115}$$

In practice very large retarding forces cannot be applied owing to the danger of deforming the incoming wire.

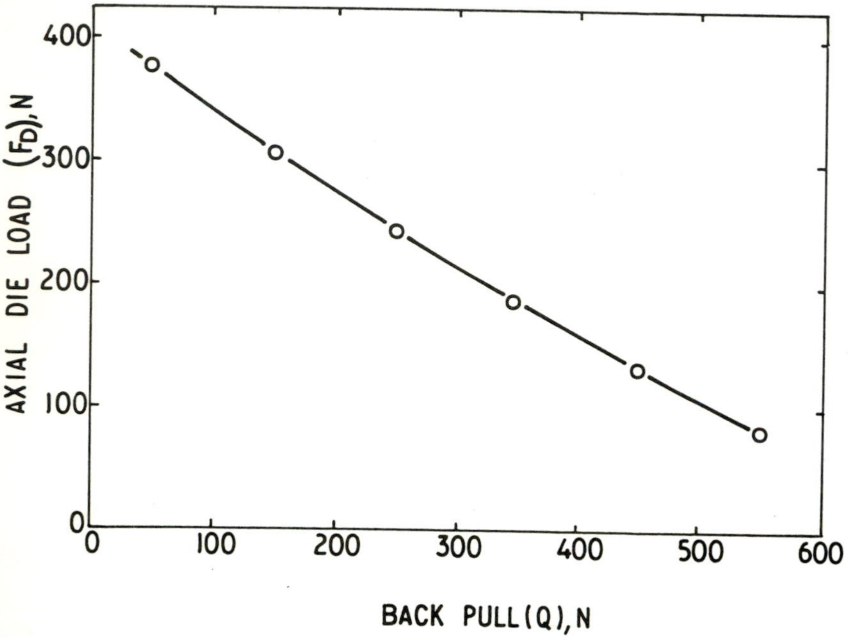

AXIAL DIE LOAD (F_D), N
400
300
200
100
0
0
100
200
300
400
500
600
BACK PULL (Q), N

Fig. 6.1

6.6 Estimation of Redundant Work and Redundant Deformation

Method (a)

Whilst a value of redundant work can be determined based on the measurement of drawing stress, there are other ways of estimating the value of ϕ_1 that are based on the displacement of flow stress curves after tensile testing the drawn wire. It has been pointed out that these methods lead to a determination of *redundant deformation*, ϕ_2, from which the redundant work factor must be calculated from the expression (Atkins and Caddell 1968)

$$\phi_1 = \left[\frac{1 - \exp(-\phi_2 B\varepsilon_w)}{1 - \exp(-B\varepsilon_w)}\right]\phi_2^n \qquad (6.15)$$

If the product $\left(B\varepsilon_w\right) \approx \frac{4\mu}{\Delta}$ (6.16)

is sufficiently small then, $\phi_1 = \phi_2^{n+1}$ (6.17)

where n is the work hardening exponent for a material behaving according to

$$\sigma = K\varepsilon^n \qquad (1.37)$$

ϕ_2 has been termed "Redundant Deformation" to distinguish it from ϕ_1. Most of the literature fails to emphasise that there is a difference. ϕ_2 is a measure of the *average* extra deformation (or equivalent strain ε^*) sustained by the rod or wire compared to the nominal (or homogeneous) strain in the wire drawing pass, ε_w. It is therefore defined as

$$\varepsilon^* = \phi_2\varepsilon_w \qquad (6.18)$$

Experimental determination of ϕ_2 has shown that it may be related to the geometry of drawing through a simple linear expression similar to equation (6.8).

$$\phi_2 = C_1 + C_2\,\Delta \qquad (6.19)$$

where Δ has been defined by equation (6.9) and C_1 and C_2 are empirically determined constants. However, experiments have shown that for several metals with widely differing work hardening characteristics (commercially pure aluminium, Armco-iron, age-hardened aluminium alloy and austenitic stainless steel) (Atkins and Caddell 1968), C_1 and C_2 can be estimated from the relationships:

$$C_1 = 2.25\ n^{0.28}\ K^{-0.10} \qquad (6.20a)$$

$$C_2 = 0.367\ n^{0.76}\ K^{-0.054} \qquad (6.20b)$$

provided that K is expressed in N/mm^2.

Values of ϕ_2 determined from these equations are accurate only to about ± 5% (Caddell and Atkins 1968) and therefore the numerical data can only be used with confidence to two significant figures. However, the determination of ϕ_2 is considerably simplified once the work hardening characteristics of the material are known. The effect of friction on ϕ_2 is thought by Atkins and Caddell only to be of secondary importance.

Example (a) - The stress-strain behaviour of annealed brass wire obeys the relationship $\sigma = K\varepsilon^n$ where $K = 1000\ N/mm^2$ and $n = 0.46$. It is drawn through a conical die of semi-angle 8° to a strain of 0.14. *Assuming* that equations (6.20a and 6.20b) hold true for brass, determine the Redundant Work factor ϕ_1 when the coefficient of friction is 0.1.

The factors in equation (6.19) can be immediately obtained from equations (6.20a and 6.20b) and from equation (6.9)

$$C_1 = 2.25\ (0.46)^{0.28}(1000)^{-0.10}$$

$$C_1 = 0.907$$

Also

$$C_2 = 0.367\ (0.46)^{0.76}(1000)^{-0.054}$$

$$C_2 = 0.140$$

and

$$\Delta \approx \frac{4\alpha}{\varepsilon} = \frac{4 \times 0.1396}{0.14} \qquad (8^o = 0.1396\ \text{rad})$$

$$\Delta = 3.989$$

Thus in (6.19)

$$\phi_2 = \underline{1.466}$$

With the coefficient of friction 0.1

$$B\varepsilon_w = 0.1 \cot 8^{\circ} \times 0.14 = 0.996$$

and substituting in (6.15) gives

$$\phi_1 = \left[\frac{1 - \exp(-1.466 \times 0.996)}{1 - \exp - 0.996}\right] 1.466^{0.46}$$

$$\phi_1 = \frac{0.1359}{0.0948} \times 1.1924 = \underline{1.71}$$

The approximation, equation (6.17), which is independent of μ and depends only on the material and the drawing geometry, gives

$$\phi_1 = 1.466^{(1.46)} = \underline{1.75}$$

With this value of μ it can be seen that the approximation gives reasonable agreement, and is therefore satisfactory for most practical drawing conditions.

Method (b)

Since redundant deformation is defined by equation (6.18) a method of finding ϕ_2 has been proposed that measures the equivalent strain ε^*, and compares it with the strain sustained in the wire drawing pass if this were all homogeneous.

If the stress-strain curves of wires drawn under different conditions are compared with the stress-strain curve of the fully annealed metal, the initial flow stress following any given strain in wire drawing is greater than the flow stress produced by straining homogeneously to the same strain in simple tension. The difference is due to redundant deformation. The strain in simple tension corresponding to the yield stress of the drawn wire is thus the equivalent strain ε^*. For a number of metals (Caddell and Atkins 1968) it has been shown that ε^* is linearly related to the nominal strain ε_w at constant die angle by

$$\varepsilon^* = K_1 \varepsilon_w + K_2 \sin\alpha \qquad (6.21)$$

and therefore from (6.18)

$$\phi_2 = K_1 + K_2 \frac{\sin\alpha}{\varepsilon_w} \qquad (6.22)$$

where K_1 and K_2 are experimentally determined constants.

A plot of ε^* against ε_w has a gradient of K_1 and an intercept on the ε^* axis of $K_2\sin\alpha$. The significance of this is that only two data points are required for one particular value of α to enable K_1 and K_2 to be found, and hence ϕ_2 can be calculated for all die angles and reductions.

It will be observed that equations (6.22) and (6.19) are identical in form, and it can be seen, equation (6.9), that $C_1 = K_1$ and $4C_2 = K_2$ over all practical wire drawing reductions.

Example (b) - For the brass of the previous example, the annealed wire is drawn to strains of 0.14 and 0.35 through conical dies with semi-angle 8°. The resulting yield (flow) stresses after these reductions are respectively, 477 N/mm^2 and 690 N/mm^2.

(i) Determine the redundant deformation factor ϕ_2, for the strain of 0.14.

(ii) Provided that the lubrication conditions remain the same, calculate the redundant deformation factor if the same material were drawn through a die of semi-angle 19° to a strain of 0.25.

(i) For a wire drawing strain of 0.14, and flow stress 477 N/mm^2 the equivalent strain is obtained from equation (1.37)

$$\sigma = 477 = 1000(\varepsilon^*)^{0.46}$$

$$\therefore \quad \varepsilon^* = 0.200$$

Thus equation (6.18)

$$\phi_2 = \frac{0.200}{0.140} = \underline{1.43}$$

(ii) In order to obtain the result for a different angle, then K_1 and K_2 need to be found. This can be done either algebraically or graphically.

As in part (i) the equivalent strain after drawing to a nominal strain of 0.35 is $\varepsilon^* = 0.446$.

We thus have:

$$\varepsilon_1^* = K_1 \, \varepsilon_{w_1} + K_2 \sin\alpha$$

and $$\varepsilon_2^* = K_1 \, \varepsilon_{w_2} + K_2 \sin\alpha$$

or $$\varepsilon_2^* - \varepsilon_1^* = K_1(\varepsilon_{w_2} - \varepsilon_{w_1})$$

$$(0.446 - 0.200) = K_1(0.35 - 0.14)$$

$$\therefore \quad K_1 = \underline{1.17}$$

$$K_2 \sin\alpha = \varepsilon_2^* - K_1 \, \varepsilon_{w_2}$$

$$\therefore \quad K_2 = \frac{0.446 - 1.17 \times 0.35}{\sin 8^{\circ}}$$

$$K_2 = \underline{0.262}$$

Thus in equation (6.22)

$$\phi_2 = 1.17 + \frac{0.262 \times \sin 19^{\circ}}{0.25}$$

$$\phi_2 = \underline{1.51}$$

The relative values of wire drawing strain and equivalent strain are shown plotted for the conditions discussed in figure 6.2.

Method (c)

The literature has many instances of redundant work being measured by the "area under the curve method" (Baron and Thompson 1951). However, the work by Trozera (1964) on internal stresses must cast some doubt on the validity of the technique, since the first assumption of Baron and Thompson that the "redundant strain in drawn wire causes an increase in strain hardening equal to that due to the same strain in simple tension" would not be fulfilled.

An alternative method has been developed that relies on measuring hardness across the transverse section of the drawn wire and relating this to strain. This is achieved by comparing the hardness with that produced in the same material after straining homogeneously in simple tension. The hardness at each point in the wire can hence be converted to an equivalent strain at that point. Graphical integration then allows the equivalent strain for the whole cross section to be found ($\bar{\varepsilon}^*$). The ratio of $\bar{\varepsilon}^*$ to nominal wire drawing strain yields a value of ϕ_2.

Since the individual hardness measurements are local then they record the effective strain at a local point in the section. Ideally Knoop hardnesses should be used since there may be a large hardness gradient between adjacent regions.

An empirical relationship

$$\varepsilon = A(HV)^m \tag{6.23}$$

has been found for annealed copper (Burke 1968) and for commercial purity aluminium (Campos et al 1980) that can be used conveniently for conversion of hardness to true strain.

Example (c) - Commercial purity aluminium is drawn from 8.75 mm to 8.00 mm in a single pass. After drawing, the wire is sectioned and the Vickers hardness measured at different depths within the specimen. The results are tabulated below:-

Distance from one surface (mm)	0.1	1	2	3	4	5	6	7	7.9
Vickers Hardness (HV)	71	66	61	58	55	60	63	65	68

Uniaxial tensile tests are conducted to different true strains on the same material, and the corresponding hardness measured. The data are shown below:

True Strain	0.02	0.05	0.10	0.17	0.30	0.40	0.50
Vickers Hardness (HV)	44.3	50.5	55.7	60.1	65.2	67.9	70.1

Determine the redundant deformation factor ϕ_2 for this reduction.

It is first necessary to plot the relationship between true strain and hardness. Figure 6.3 shows the results on a log-log scale and in this case it gives a good straight line. The gradient of the line gives the exponent m, and A can then be immediately calculated from the equation.

$$m = 7.0 \text{ and } A = 6 \times 10^{-14} \ (\text{kgf/mm}^2)^{-7}$$

or

$$\varepsilon = \underline{6 \times 10^{-14}(\text{HV})^7}$$

Using the equation (or the graph) the equivalent true strain ε^* for each value of hardness across the section can be found. These are shown below, and are plotted in figure 6.4.

Hardness	71	66	61	58	55
Equivalent strain ε^*	0.545	0.327	0.189	0.133	0.091

Hardness	60	63	65	68
Equivalent strain ε^*	0.168	0.236	0.294	0.403

The average equivalent strain can be obtained by summing the strains over the whole area of the wire.

$$\text{Thus } \bar{\varepsilon}^* = \frac{1}{A}\int \varepsilon^* dA. \sim \frac{1}{A}\sum \varepsilon^* \ 2\pi a da$$

where dA is a small increment of area and da is a small increment of radius.

Taking da as ½ mm and reading from both sides of the centre-line in figure 6.4 values of a and ε* can be read off as follows:

Interval	1	2	3	4
a (mm)	0.25	0.75	1.25	1.75
ε* %	$\frac{11+10.5}{2}$	$\frac{14.5+12.5}{2}$	$\frac{18.0+14.5}{2}$	$\frac{21.5+16.5}{2}$
ε* ada	1.34	5.06	10.16	16.63

Interval	5	6	7	8
a (mm)	2.25	2.75	3.25	3.75
ε* %	$\frac{24.5+21.5}{2}$	$\frac{27.5+29}{2}$	$\frac{32.5+38}{2}$	$\frac{39+51}{2}$
ε* ada	25.88	38.84	57.28	84.38

$$\sum = 239.6 \times 2\pi$$

$$\therefore \quad \bar{\varepsilon}^* = \frac{1}{\pi(4)^2} \cdot 2\pi \sum \varepsilon^* ada = \frac{239.6}{8} = 29.95\% = \underline{0.300}$$

If the exit diameter is 8 mm and inlet diameter = 8.75 then

$$\varepsilon_w = 2\ell n \frac{D_o}{D_f} = 0.179$$

Substituting in equation (6.18)

$$\phi_2 = \frac{0.300}{0.179} = \underline{1.68}$$

Although this method appears to be very satisfactory for a single pass, Campos et al (1980) have shown that extending the technique to multiple passes is much less satisfactory. The authors were unable to explain why the values of ϕ_2 that they obtained in multipass work were significantly lower than expected from theory, and even included values less than unity.

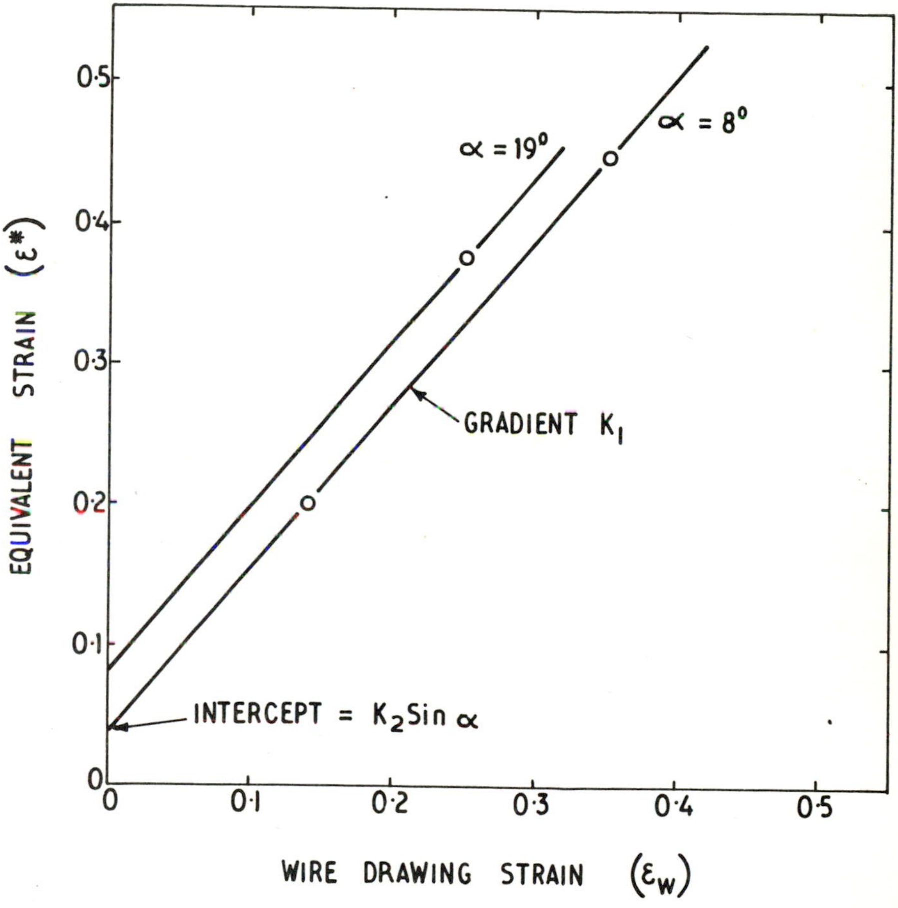

EQUIVALENT STRAIN (ε^*)
0·5
0·4
0·3
0·2
0·1
0
α = 19°
α = 8°
GRADIENT K_1
INTERCEPT = $K_2 \sin \alpha$
0
0·1
0·2
0·3
0·4
0·5
WIRE DRAWING STRAIN (ε_W)

Fig. 6.2

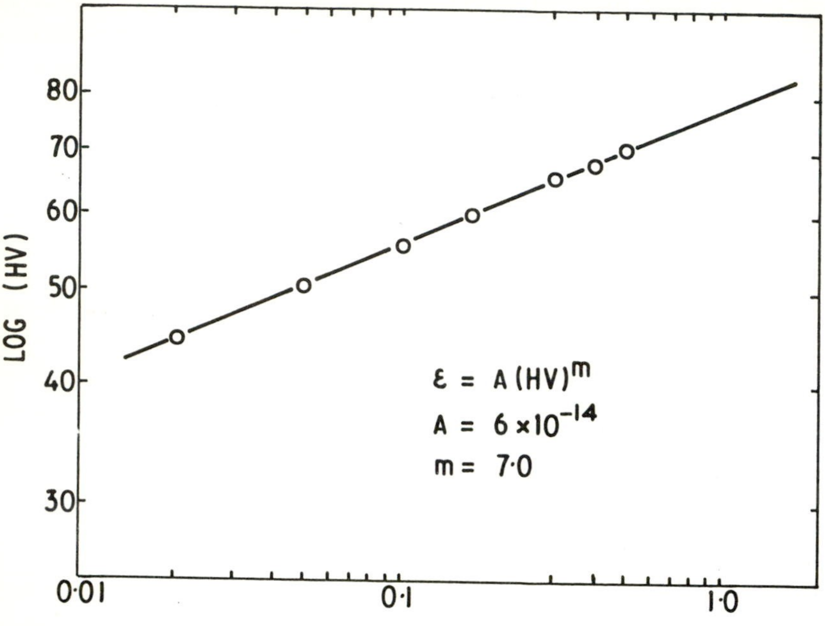

Fig. 6.3 LOG (TRUE STRAIN)

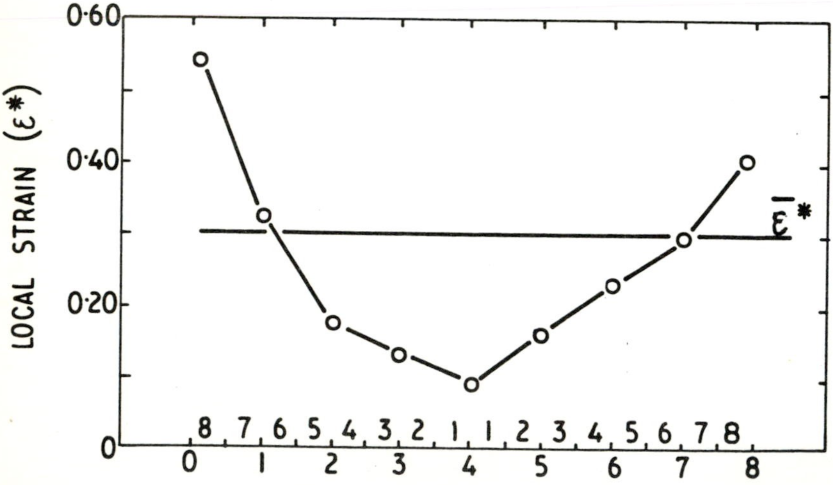

Fig. 6.4 DISTANCE FROM ONE SURFACE, mm.

6.7 Limiting Reduction in Wire Drawing

(a) Non-Hardening

When wire is drawn through a die, the required stress that must be applied is always less than the tensile flow stress of the same material deformed uniaxially to the same strain as the wire. This is due to the assistance given to the drawing stress by the reactive compressive stresses in the die. As the strain per pass increases then the drawing stress more nearly approaches the tensile flow stress. When the strain is such that the two are coincident, then the wire will fail at the die exit due to simple tensile fracture of the drawn metal.

Since the mean flow stress of the material is constant for a non-hardening material then at the limiting reduction equation (6.3) can be written (Rowe 1965):

$$\frac{\sigma_d}{\bar{\sigma}} = 1 = \beta \, \ell n \frac{1}{1-r_{lim}} \qquad (6.24)$$

In the absence of friction and redundant work when $\beta = 1$.

$$\frac{1}{1-r_{lim}} = \exp(1.0)$$

$$\text{or } r_{lim} = \underline{0.63}$$

Considerations of friction and redundant work using equation (6.4) might also be included in the determination of the strain limit. Using the same method as above, the expression for the maximum reduction can be obtained by rearrangement of equation (6.4) with the average flow stress and the drawing stress again being made equal. Therefore,

$$r_{lim} = 1 - \exp\left\{\frac{\alpha(2\alpha - 3)}{3(\alpha + \mu)}\right\} \qquad (6.25)$$

Example - Determine the maximum drawing reduction for brass, in the presence of friction and redundant work, using the same conditions as example (a) in section 6.6.

Since $\alpha = 8^{\circ} = 0.1396$ rad

and $\mu = 0.1$

then $$r_{lim} = 1 - \exp\left\{\frac{0.1396(2 \times 0.1396 - 3)}{3(0.1396 + 0.1)}\right\}$$

or $r_{lim} = \underline{0.41}$

It is evident that the strain is *reduced* by taking friction and redundant work into account compared to the case of perfect lubrication and homogeneous straining. This is clearly because the drawing stress rises much more quickly with strain in the former case and thus reaches the maximum stress at a lower strain. Calculation of the different contributions to the increased drawing stress from equation (6.4) shows that, for the conditions under consideration, the friction has a much greater influence than the redundant work.

(b) With Work Hardening

Not only is the maximum reduction controlled by the material being drawn, it is also dependent on the condition of the "tag" end that is gripped by the jaws of the machine. This tag end may be in the annealed condition in which case it will determine the limiting drawing stress that can be sustained, since this limit will be the tensile strength of the tag, Caddell and Atkins (1968). Although the limiting reductions calculated are realistic, the equations involved are not simple and involve numerical solutions, and are therefore not considered here. Additionally, the tag end is often swaged, or work hardened, prior to insertion into the die hole. Under these conditions the maximum reduction occurs when the stress needed to draw the wire reaches the value of the yield strength of the drawn wire as before. The tag at all times is considered to be capable of carrying the loads. In the presence of both work hardening and redundant work the strain limit is given by:

$$\left(\phi_2 \varepsilon_w\right) = \frac{1}{B} \ell n \frac{(1 + B)}{(1 - nB)} \qquad (6.26)$$

and thus according to equation (6.22) we can write

$$\varepsilon_w = \frac{1}{C_1 B} \ell n \frac{(1 + B)}{(1 - nB)} - \frac{4C_2 \sin\alpha}{C_1} \qquad (6.27)$$

Depending on the precise values of μ and α chosen the drawing limit may be greater or less than the value calculated on the basis of homogeneous deformation. The reason for this lies in the fact that the final flow stress of the wire at exit is considered in the calculation rather than the average flow stress of the pass, which is rather less. A second factor that leads to increased limiting strain is possibly the presence of redundant work. This has the effect of work hardening the wire so that the effective strain is much greater than the measured drawing strain. Thus the wire can sustain greater loads due to both these reasons.

Example - For the previously considered annealed brass, section 6.6, find the maximum drawing reduction if the coefficient of friction were 0.1.

The terms required are:

$B = \mu\cot\alpha = 0.1 \tan 82^{\circ} = 0.7115$

$\sin\alpha = 0.1392$

The work hardening coefficient = 0.46

and $C_1 = 0.907$ and $C_2 = 0.140$

Thus

$$\varepsilon_w = \frac{\ell n \left[\dfrac{1.7115}{1-0.46 \times 0.7115}\right]}{0.907 \times 0.7115} \qquad \frac{4 \times 0.140 \times 0.1392}{0.907}$$

$$\varepsilon_w = \underline{1.36} \text{ and } r_{lim} = \underline{0.74}$$

At this large reduction reference to equation (6.19) shows that the factor ϕ_2 is less than unity. This is clearly impossible and the redundant work equations no longer hold at such small values of Δ (Caddell and Atkins 1968). The limiting value of ϕ_2 is therefore 1.0; it can thus be deduced that there is no redundant work for this high reduction and the increased wire drawing strain over the average yield stress case is due to the homogeneous work hardening of the metal.

These very high reductions are never achieved in practice since other factors become dominant, such as the breakdown of the lubricant film leading to metal "pick up" on the die surface. For most practical purposes the limiting reduction is of the order of 35-45% (Rowe 1965).

6.8 <u>Tensile Strength of the Metal after Drawing</u>

The extensive analysis of redundant work carried out by Caddell and Atkins (1968) has facilitated the determination of the strength properties following wire drawing. The equations proposed assume that equation (1.37) holds for the metal and that the metal is *uniformly* strain hardened. The tensile strength S_w is obtained by simple modification of standard equations, replacing the strain in the equations by the equivalent strain. As the wire drawing strain increases, then the yield stress of the subsequently tested wire approaches the tensile strength of the material. Eventually the drawing strain is so large that on subsequent tensile testing necking occurs in the metal as soon as the tensile yield stress is reached. These two conditions may be expressed as:

$$S_w = S \exp(\phi_2 \varepsilon_w) \qquad (6.28)$$

$$\text{when } \phi_2 \varepsilon_w \leqslant n$$

and S is the uniaxial tensile strength of the fully annealed material.

$$S_w = K(\phi_2 \varepsilon_w)^n \qquad (6.29)$$

$$\text{when } \phi_2 \varepsilon_w \geqslant n$$

This latter condition applies where the equivalent strain induced by wire drawing exceeds the strain at the onset of necking in simple tension.

These equations have also been tested for a number of different metals and give useful agreement with the experimentally measured values of tensile strength.

However, as has already been mentioned, Campos et al (1980) showed that the values of ϕ_2 determined in a single die pass and in multiple die passes to the same total strain are not the same. These equations therefore are strictly only applicable to strain produced in a single pass through conical dies. Nevertheless, they may well give useful comparisons under differing drawing conditions, provided that these limitations are borne in mind.

<u>Example</u> - In example (b,ii), section 6.6, the redundant deformation factor of annealed brass was shown to be 1.51 for a wire drawing strain of 0.25. Find the tensile strength of the drawn wire.

$$\phi_2 = 1.51 \text{ when } \varepsilon_w = 0.25$$

The tensile strength, S, of the fully annealed metal corresponds to the stress when the true strain is numerically equal to the work hardening coefficient

$$\text{i.e.} \quad \text{when } \varepsilon = n = 0.46$$

The product $\phi_2\varepsilon_w = 0.378 < 0.46$, hence the first equation (6.28) should be used since necking does not occur in the wire at the yield stress.

The true stress at the onset of necking is, from equation (1.37)

$$\sigma = 1000(0.46)^{0.46}$$

$$\sigma = 699.6 \text{ N/mm}^2$$

From equation (1.9)

$$S = \sigma \frac{A}{A_o} = \sigma \exp(-\varepsilon)$$

$$= 699.6 \exp(-0.46)$$

$$S = 441.6 \text{ N/mm}^2$$

Thus in equation (6.28)

$$S_w = 441.6 \exp(1.51 \times 0.25)$$

$$S_w = \underline{644.1 \text{ N/mm}^2}$$

It can be seen that the tensile strength following prior cold work is increased in comparison with the annealed metal. This method of work hardening a metal by some compressive type of deformation process in order to improve the strength properties is, of course, industrially very important since it greatly extends the useful range of applications.

6.9 Theoretical Power Consumption

Interest in drawing loads and in the factors that affect them stems from the desirability of operating the drawing equipment with the minimum power consumption. In addition it is advantageous that the correct size of motor is selected to make the desired reductions.

The motive force for drawing is supplied through bull blocks and capstans; their speed and size depending on the size of rod or wire being drawn. The minimum power required is given by the rate at which work is performed:

$$\text{Power} = \text{Drawing Load} \times \text{Speed of Drawing} = F \times v$$

$$= A_f\ \bar{\sigma}\ \varepsilon_w\ g(\alpha,\mu) \times v \qquad (6.30)$$

where v is the drawing speed and $g(\alpha,\mu)$ is a function of the wire drawing parameters, and is obtained from equations 6.4 - 6.7.

Two of these equations have been used extensively by Duckfield (1973) in practical schedule calculations, a modified Sachs equation:

$$g(\alpha,\mu) = \phi_1\ (1 + \mu\cot\alpha)$$

$$\approx A' \left\{1 + \frac{\mu}{\alpha} + \frac{C'(\alpha+\mu)}{A'\ \varepsilon}\right\} \qquad (6.31a)$$

(where A' and C' are constants that have been determined experimentally by Johnson and Rowe (1968) as 0.88 and 0.78 respectively) and Siebel's equation:

$$g(\alpha,\mu) = \left(1 + \frac{\mu}{\alpha} + \frac{2\alpha}{3\varepsilon}\right) \qquad (6.31b)$$

Values of $g(\alpha,\mu)$ determined over a range of die angles and drawing strains using a typical published value of $\mu = 0.05$ lie above about 1.5 irrespective of the method of finding $g(\alpha,\mu)$. Thus for many problems it is convenient to use this minimum value (Duckfield 1973).

If the wire, diameter d, is drawn on a capstan, diameter D, rotating at N revolutions per second then the linear drawing speed is:

$$v = \pi(D+d)N \qquad (6.32a)$$

$$v \sim \pi DN \qquad (6.32b)$$

since for many commercial machines D >> d.

<u>Example</u> - A bull block, 850 mm diameter, rotates at 8 rpm and draws carbon steel rod through a conical die of semi-angle 8°. The drawing conditions are such that the redundant work factor is 1.18 and the coefficient of friction is 0.05. Ignoring any mechanical losses in the system, determine whether or not it is possible to draw 25 mm diameter rod to 22 mm diameter in a single pass without overload if the block is driven by a motor rated at 40 horsepower. Assume the average flow stress of the steel is 600 MN/m^2.
(Many motors in Britain are still rated in hp, and 1 hp ≡ 746 W)

Using equation (6.31a)

$$g(\alpha,\mu) = 1.18(1 + 0.05 \cot 8^{o}) = 1.60$$

From equation (6.32a) $v = \pi(0.85 + 0.022)\,\frac{8}{60} = 0.365$ m/s

$$\text{Thus power required} = \frac{\pi(22)^2}{4} \times 10^{-6} \times (600 \times 10^{6}) \times \ell n\left(\frac{25}{22}\right)^2 \times 1.60 \times 0.365$$

$$= 34054.\ \text{W}$$

$$\therefore \text{ hp developed} = \frac{34054}{746} = \underline{45.7}$$

It is therefore clear that this reduction in area could not be made without overload.

6.10 Maximum Drawing Speed and Output

Work conducted on heavy carbon steel wire over a wide range of sizes, strengths, reductions, speeds, coatings and lubricants has shown that there is a linear relationship between the power consumed by the drive motors and the power calculated theoretically (Duckfield 1973).

$$\text{Measured Power} = M\ (\text{Theoretical Power}) + C$$

$$= M\ (A_f\ \bar{\sigma}\ \varepsilon_w\ g(\alpha,\mu)v) + C \qquad (6.33)$$

Both M and C are constants that can be determined graphically by measuring the required power and comparing this with the value calculated from equation (6.30).

M is the gradient of the line and involves the mechanical efficiency and C is the intercept at zero calculated power, i.e. is the power needed to run the machine when empty.

Where a single drawing machine contains several die blocks the power required at each block can be summed to give the total for the system. Although the terms in the bracket in equation (6.33) vary from die to die, they can be satisfactorily evaluated overall using the mean values of flow stress and strain for the whole pass sequence. A_f and v then refer to the wire after the final pass.

Once M and C are known for the machines in question then the maximum drawing speed v and output can be determined from equation (6.33). The information is very useful in planning and costing a production run.

$$\text{Weight/hr} = (\text{vol/hr}) \times \text{density}$$

$$\text{Weight/unit time} = A_f\ v\ \rho \qquad (6.34)$$

Example - Following a prior strain to 0.5 a 0.8% carbon steel wire is reduced on a six hole drawing machine from 6 mm to 3 mm diameter. The total motor capacity is rated at 400 kW and the machine obeys the empirical relationship:

Measured Power = 2.0 (Theoretical Power) + 40

where power is in kW.

If the steel obeys the relation for flow stress change during wire drawing:

$$\sigma = K\varepsilon_w^n \text{ where } K = 1600 \text{ N/mm}^2$$

$$\text{and} \quad n = 0.265$$

Determine the drawing speed at the last die and the output per hour, neglecting handling time. Density of steel = 7.8 Mg/m^3.

Equation (6.33),

$$\text{Calculated Power} = \frac{400 - 40}{2.0} = 180 \text{ kW}$$

$$\text{Total strain in wire } \varepsilon_w = \ell n\left(\frac{6}{3}\right)^2 = 1.386$$

Average flow stress for the pass sequence is obtained as in equation (1.38),

$$\therefore \quad \bar{\sigma} = \frac{1}{\varepsilon_1 - \varepsilon_0}\int_{\varepsilon_0}^{\varepsilon_1} K\left[\varepsilon_w\right]^n d\varepsilon$$

$$\text{i.e.} \quad \bar{\sigma} = \frac{K}{(\varepsilon_1 - \varepsilon_0)(n+1)}\left\{\varepsilon_1^{\,n+1} - \varepsilon_0^{\,n+1}\right\}$$

where $\varepsilon_0 = 0.5$ and $\varepsilon_1 = 0.5 + 1.386 = 1.886$

Thus $\bar{\sigma} = 1.657 \times 10^9$ N/m^2.

Using this value of flow stress and the minimum value of $g(\alpha,\mu)$ of 1.5 discussed in the previous section, the speed of drawing from equation (6.33) is given by

$$v = \frac{180 \times 10^3}{1.5 \times 1.657 \times 10^9} \times 1.386 \times \pi\frac{3^2}{4} \times 10^{-6}$$

$$v = \underline{8.53 \text{ m/s}}$$

Also from equation (6.34),

$$\text{Wt drawn/hr} = \pi\frac{3^2}{4} \times 10^{-6} \times 8.53 \times 7.8 \times 10^3 \times 3600 \times 10^{-3} \text{ tonne/hr}$$

$$\text{Output} = \underline{1.69 \text{ tonne/hr}}$$

6.11 Draft Sequence in Wire Drawing

When material is drawn through a succession of dies the work done on the stock will both increase the internal energy in the wire, i.e. work hardening will occur, and cause deformational heating due to the combined effects of homogeneous work, friction, and redundant work. Work hardening leads to progressively larger amounts of work being done on the wire if the strain sustained in each pass is constant. In addition the temperature rise in later passes can be substantial and may lead to problems of ageing of the wire or to lubricant breakdown. It has, therefore, been pointed out (Duckfield 1971) that to prevent ageing effects maximum cooling ought to be on the last pass and immediately afterwards.

If a uniform drafting is maintained throughout the die sequence then the last die will reach its maximum output before the others reach their maximum power development. This leads to a lower than necessary productivity and the earlier drawing machines will operate at reduced efficiency. These problems can be overcome if the drawing machine is operated so that the workload is uniform over all die blocks.

The calculations necessary to ensure uniform drafting for any number of dies have been presented by Duckfield (1973).

The basic equation for the work load per unit volume is given by:

$$\left[1.12 + \frac{0.085}{\varepsilon_w}\right] (\bar{S} \times \varepsilon_w) = \text{Constant} \qquad (6.35)$$

where $\bar{S}$ is the average value of the metal's tensile strength over the whole pass. Duckfield considers this to be a more useful parameter to use than flow stress since wire tensile strengths are commonly measured in commercial drawing practice.

The term in the first bracket is related to $g(\alpha,\mu)$ discussed previously. It was empirically determined by Rigo (1972) on steel wire at production drawing speeds over a wide range of drawing conditions but with only one lubricant.

In order to make calculations for a number of dies a small computer is ideally suited. In this circumstance assumptions that are convenient for ease of calculations by hand are not necessary, and the labour is significantly reduced.

A reasonable assumption is that the tensile strength of the wire changes linearly with wire drawing strain

$$S = S_o + m\varepsilon \qquad (6.36)$$

where m is a constant. In fully annealed and high tensile steel wire this relationship is not strictly true. However, it can be dealt with by dividing the actual curve into straight line segments satisfying the above equation with different values of the exponent m (Duckfield 1973).

The second assumption is that the mean true strain per die ε_n is approximately constant. Although for taper drafting this is not strictly true the taper is relatively shallow and this allows simplification of a complex mathematical expression. Examination of the subsequent calculations on drafting seem to indicate that the approximation does not lead to serious error.

With these two assumptions it can be shown that (Duckfield 1973):-

$$w_T = 1.12\ \bar{S} \cdot (\varepsilon_T + 0.076\ n) \qquad (6.37a)$$

and $\therefore$

$$w_n = 1.12\ \bar{S} \cdot (\varepsilon_n + 0.076) \qquad (6.37b)$$

where w_T and w_n are respectively the total work per unit volume over all dies and the work per unit volume at the nth die. $\bar{S}$ is the average overall tensile strength of the wire over n pass sequences. The average workload on each die can thus be determined.

The necessary reduction to give this workload can then be found sequentially from the first to the nth die taking work hardening into account, by application of the further relationships:

$$\varepsilon_n = x_n\left\{1 - \frac{m}{2S_{in}} \; x_n\right\} \tag{6.38}$$

where

$$x_n = \left\{\frac{w_n}{1.12\ S_{in}} - 0.076\right\} \tag{6.39}$$

where S_{in} is the inlet tensile strength of the wire at the nth die. Once the required strain in the nth die is known the die size can be calculated and hence the tensile strength of the material on exit. This is then the value of $S_{i(n+1)}$. The values for the whole balanced die set can thus be determined very rapidly. The repetitious nature of the calculations makes a small programmable calculator of great value.

Example - 0.8% C steel wire is drawn in six passes from 6 mm to 3 mm diameter. The initial patented tensile strength of the wire is 1330 N/mm² and the work hardening rate is 400 N/mm² per unit strain. Determine the reductions for each drawing die necessary to ensure a uniform workload per die block.

$$\text{Total Strain } \varepsilon_T = \ell n \frac{6^2}{3^2} = 1.386$$

Therefore the final tensile strength from equation (6.36) is

$$S = 1330 + (400 \times 1.386) = 1884 \text{ N/mm}^2$$

and the overall mean tensile strength is

$$\bar{S} = \tfrac{1}{2}(1884 + 1330) = 1607 \text{ N/mm}^2$$

The approximate workload per die is obtained from equation (6.37b) using the average strain per die.

$$w_n = 1.12 \times 1607 \left(\frac{1.368}{6} + 0.076\right)$$

$$w_n = 552.5 \text{ N/mm}^2 \text{ per die } (\equiv \text{MJ/m}^3)$$

Consider the <u>First Die</u>:

In equation (6.39)

$$x_1 = \left\{\frac{552.5}{1.12 \times 1330} - 0.076\right\} = 0.295$$

and in equation (6.38)

$$\varepsilon_1 = 0.295 \left\{1 - \frac{400}{2 \times 1330} \times 0.295\right\}$$

$$\varepsilon_1 = \underline{0.282}$$

Now since the inlet size is 6 mm, then the exit size from equation (1.4) is 5.21 mm. This also, from equation 6.36, gives a work hardened (drawn) strength of :

$$S_{i2} = 1330 + 400\ (.282) = 1443\ \text{N/mm}^2$$

The true workload per unit volume for this die, w_1, can be found from equation 6.37b:

$$w_1 = 1.12 \left(\frac{1443 + 1330}{2}\right)(0.282 + 0.076)$$

$$w_1 = \underline{555.8\ \text{MJ/m}^3}$$

Now consider the <u>Second Die</u>:

The new value of $S_i = 1443\ \text{N/mm}^2$
and hence the value of

$$x_2 = \left(\frac{552.5}{1.12 \times 1443} - 0.076\right) = 0.266$$

This gives $\varepsilon_2 = 0.266 \left\{1 - \frac{400}{2 \times 1443} \times 0.266\right\}$

$$\varepsilon_2 = \underline{0.256}$$

Thus the wire exit diameter is 4.58 mm and the tensile strength on exit is 1443 + 400 x 0.256 = 1545 N/mm^2.

The true workload per unit volume at this die should be virtually the same as at the previous die.

$$w_2 = 1.12 \left\{\frac{1545 + 1443}{2}\right\} (0.256 + 0.076)$$

$$w_2 = \underline{555.6 \text{ MJ/m}^3}$$

This procedure can be followed for each successive die. The calculated data for all six dies are laid out in the following table and plotted in figure 6.5.

Table 6.2

	TAPER DRAFTING					EQUAL DRAFTING	
Die No	ε	Final Dia.mm	Reduction in Area %	S_i	w	ε	w
1	0.282	5.21	24.6	1443	555.8	0.231	473.1
2	0.256	4.58	22.6	1545	555.6	0.231	504.8
3	0.235	4.07	20.9	1639	554.5	0.231	536.4
4	0.219	3.65	19.7	1727	556.0	0.231	568.0
5	0.205	3.30	18.5	1809	556.4	0.231	599.7
6	0.193	3.00	17.6	1886	556.6	0.231	631.3

(In the figure the workload per die has also been presented on the basis of equal strain per die).

Reference to equation (6.33) in the previous section shows that as the workload on a particular die increases then, for a fixed maximum power output on the drive motor, the maximum speed of drawing decreases. Consideration of the final draft reveals that by reducing the workload per unit volume from 631.3 to 556.6 MJ/m^3 the output from the machine can be increased by over 13%. Alternatively a reduction in the workload will lead to a useful reduction in heat generated, section 1.7. In fact it is better that the last draft should be even less than the equal workload value to avoid the problem of heating.

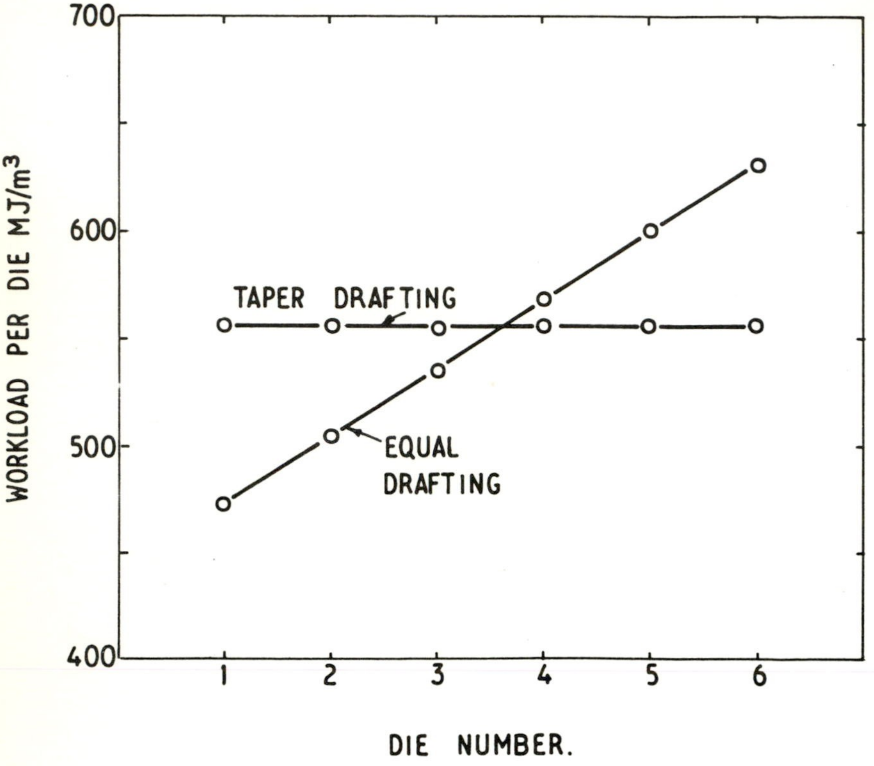
700
600
500
400
WORKLOAD PER DIE MJ/m³
TAPER DRAFTING
EQUAL
DRAFTING
1
2
3
4
5
6
DIE NUMBER.

Fig. 6.5

BIBLIOGRAPHY

Alexander, J.M. (1954-55), J. Mech. Phys. Solids, 3, 233.
Alexander, J.M. (1982), Private Communication.
Atkins, A.G. and Caddell, R.H. (1968), Int. J. Mech. Sci., 10, 15.
Avitzur, B. (1968), "Metal Forming: Processes and Analysis", McGraw Hill Pub. Co., N.Y., U.S.A., p 173.
Avitzur, B. (1970), "Metal Forming: Interrelations between Theory and Practice", Plenum Press, N.Y., U.S.A., p 40.
Backofen, W.A. (1972), "Deformation Processing", Addison-Wesley Pub. Co., Reading, Mass., U.S.A., p 194.
Baron, H.G. and Thompson, F.C. (1950-51), J. Inst. Metals, 78, 415.
Barraclough, D.R., Whittaker, H.J., Nair, K.D. and Sellars, C.M. (1973), J. Testing and Evaluation, 1, 220.
Beese, J.G. (1972), JISI, 210, 433.
Bland, D.R. and Ford, H. (1948), Proc. Inst. Mech. Eng., 159, 144.
Burke, J.J. (1968), reported in Backofen, W.A. (1972), pp 138-139.
Caddell, R.M. and Atkins, A.G. (1968), J. of Eng. for Ind., May, 411.
Campos, E.B., Duarte, M.R. and Cetlin, P.R. (1980), Associacao Brasiliera de Metalurgia, (ABM), Conf., v3, p 63, Sao Paulo, Brazil.
Castle, A.F. and Sheppard, T. (1976), Metals Tech., 3, 465.
Cook, P.M. (1957), Proc. Conf. "The properties of materials at high rates of strain", 86-97, Instn. Mech. Eng., London.
Cook, M. and Larke, E.C. (1945), J. Inst. Metals, 71, 371.
Cook, P.M. and McCrum, A.W. (1958), "The Calculation of Load and Torque in Hot Flat Rolling", Brit. Iron and Steel Assoc., London.
Duckfield, B.J. (1971), Wire Ind., 38, 43 and 120.
Duckfield, B.J. (1973), Wire Ind., 40, 618 and 702.
Ekelund, S. (1933), Steel, 93, Nos. 8-14.
Farag, M.M. and Sellars, C.M. (1973), J. Inst. Metals, 101, 137.
Fields, D.S. and Backofen, W.A. (1957), Proc. Am. Soc. Testing Mat., 57, 1259.
Fishenden, M. and Saunders, O.A. (1934), "Calculation of Heat Transmission", H.M. Stationery Office.

Ford, H. (1948), Proc. Inst. Mech. Eng., 159, 115.

Helmi, A. and Alexander, J.M., (1969), JISI, 207, 1219.

Hill, R. (1951), see Baron, H.G. and Thompson, F.C. (1950-51).

Hill, R. and Tupper, S.J. (1948), JISI, 159, 353.

Hirst, S. and Ursell, D.U. (1958), Metal Treatment, 25, 409.

Hitchcock, J.H. (1935), "Roll Neck Bearings", ASME Res. Pub.

Hollander, F. (1970), in "Mathematical models in metallurgical process development", ISI Spec. Rep. 123, 46.

Hughes, K.E., Nair, K.D. and Sellars, C.M. (1974), Metals Tech, 1, 161.

Hughes, K.E. and Sellars, C.M. (1972), JISI, 210, 661.

Johnson, W. and Mellor, P.B. (1973), "Engineering Plasticity", Van Nostrand, Reinhold Co., London.

Johnson, R.W. and Rowe, G.W. (1968), J. Inst. Metals, 96, 97.

Kudo, H. (1960), Int. J. Mech. Sci., 2, 102.

Kuhn, H.A. and Downey, C.L. (1971), Int. J. of Powder Met., 7(1), 13.

Kuhn, H.A. and Downey, C.L. (1973), Trans. ASME, J. of Eng. Mat. and Tech., Jan., 41.

Larke, E.C. (1967), "The Rolling of Strip Sheet and Plate", Science Paperbacks Ltd.

Male, A.T. and Cockcroft, M.G. (1964-65), J. Inst. Metals, 93, 38.

McGannon, H.E. (1971), Edit., "The Making Shaping and Treating of Steel", U.S. Steel Corp., p 636.

Nadai, A. (1949), J. App. Phys., 16, 349.

Orowan, E. (1943), Proc. Inst. Mech. Eng., 150, 140.

Parkins, R.N. (1968), "Mechanical Treatment of Metals", Instn. of Metallurgists, p 63.

Rigo, J.H. (1972), Wire J., 5, 43.

Roberts, W.L. (1978), "Cold Rolling of Steels", M. Dekker Inc., N.Y.

Rowe, G.W. (1965), "Principles of Metalworking", Edward Arnold (London).

Sachs, G. (1927), Zeit Angew. Mat u. Mech., 7, 235.

Sellars, C.M. (1981), 24th Colloque de Métallurgie, Les Traitments Thermoméchaniques: Aspects Théoretiques et Applications, INSTN, Saclay, France.

Sellars, C.M. and Tegart, W.J. McG. (1972), Int. Met. Rev., 17, 1.

Sellars, C.M. and Whiteman, J.A. (1981), Metals Tech., 8, 10.

Schey, J.A. (1970), "Metal Forming. Interrelation between Theory and Practice", Plenum Press, N.Y., U.S.A., p 275.

Schroeder, W. and Webster, D.A. (1949), J. App. Mech., 16, 289.

Sheppard, T. (1981), Metals Tech. 8, 130.

Sheppard, T. and Wood, E.P. (1980), Metals Tech., 7, 58.

Siebel, E. (1947), Stahl u. Eisen, 66-67, 171.

Sims, R.B. (1954), Proc. Inst. Mech. Eng., 168, 191.

Sparling, L.G.M. (1977), Int. Mech. Rev., 22, 303.

Sykes, C. (1953-54), Trans. N.E. Coast Inst. of Eng. and Shipbuilders, 70, 6.

Tegart, W.J. McG. (1966), "Elements of Mechanical Metallurgy", Macmillan Series in Materials Science, N.Y.

Timoshenko, S.P. and Gere, J.M. (1973), "Mechanics of Materials", Van Nostrand, Reinhold Co., London.

Trozera, T.A. (1964), Trans. A.S.M., 57, 309.

Underwood, L.R. (1950), "Rolling of Metals", v1, Wiley.

Whitton, P.W. (1957-58), J. Inst. Metals, 86, 417.

Wilcox, R.J. and Whitton, P.W. (1958-59), J. Inst. Metals, 87, 289.

Wistreich, J.G. (1955), Proc. Inst. Mech. Eng., 169, 654.

Wistreich, J.G. (1958), Met. Reviews, 3, 108.